T0026457

IN OUR OWN IMAGE

BY THE SAME AUTHOR:

Novels:

The Island Survival Guide

The Passage

Archipelago Republic

End of the East

The Secrets of the Lands Without

Plays:

Y

Baby

Turing

IN OUR OWN IMAGE

>> <<

SAVIOR or DESTROYER? THE HISTORY and FUTURE of ARTIFICIAL INTELLIGENCE

GEORGE ZARKADAKIS

PEGASUS BOOKS
NEW YORK LONDON

IN OUR OWN IMAGE

Pegasus Books LLC
80 Broad Street, 5th Floor
New York, NY 10004

Copyright © 2015 by George Zarkadakis

First Pegasus Books hardcover edition March 2016

All rights reserved. No part of this book may be reproduced in whole or in part without written permission from the publisher, except by reviewers who may quote brief excerpts in connection with a review in a newspaper, magazine, or electronic publication; nor may any part of this book be reproduced, stored in a retrieval system, or transmitted in any form or by any means electronic, mechanical, photocopying, recording, or other, without written permission from the publisher.

ISBN: 978-1-60598-964-8

10 9 8 7 6 5 4 3 2 1

Printed in the United States of America
Distributed by W. W. Norton & Company, Inc.

CONTENTS

INTRODUCTION

I met my first robot when I was five. It was a warm summer evening in Athens when my mother took me to an open-air cinema close to our house. Only a month earlier the first men had landed on the Moon and the world was abuzz with space fever. Every kid wanted to become an astronaut. We would dress up in all kinds of oddments that looked vaguely space-like and pretend to bounce about in zero gravity. It was a no-brainer for the owners of the cinema to screen a film set in the distant future, where humans would casually visit stars on the other side of the galaxy.

The cinema was jam-packed. I recall the aroma of jasmine mixing with pungent popcorn and sweat, the chatter of people and of crickets, the uncomfortable chair, the excitement of going to the movies in the days when television was a luxury afforded only by a few. My mother and I were ushered to our seats as the lights dimmed. The spirited babbling of the audience ebbed. Loud music blasted out of the speakers. The dark screen flickered and became a window to another world: there was a spaceship that looked like a round pie, and men in shiny uniforms speaking a language I did not understand. As for the subtitles, I was too young to read them. Maybe it was somewhat premature of my mother to take me to that film; nevertheless, watching *Forbidden Planet* in Athens that night changed my life for ever.

I am told that in the following days I would scribble countless sketches of the one thing that had impressed me the most: Robby the Robot. I would draw him with his lights flashing, explaining to whomever had the patience to listen how they flashed when he spoke in his mechanical voice, waving his arms and running about. Without question, a mechanical thing that walked, talked – and obeyed orders – would make the ideal playmate. Moreover, with a super-strong robot following me around, who would dare bully me? I could return the favour by teaching him all the things I knew: how to kick a football through a window, how to chase cats, how to draw. We would be the best of friends, pals for life. Robby the Robot had me in his grip for days, and held me tightly therein well into my final years at school when I decided to become an engineer and build my own robot one day. And that's how my journey into Artificial Intelligence began.

And quite a journey it was, too, for I literarily had to pack my suitcases and fly to London to study at university. My choice of subject was Control and Systems Engineering, a discipline based on the theory of cybernetics as developed by the American mathematician Norbert Wiener in the 1940s.

Wiener is one of the demigods of Artificial Intelligence. Born in Missouri in 1894, he was a child prodigy who earned a degree in mathematics at the age of fourteen and a doctorate at seventeen. A polymath with an insatiable appetite for knowledge, Wiener studied philosophy as well as zoology, then travelled to Europe to learn from the most prominent mathematical celebrities of the early twentieth century: Bertrand Russell at Cambridge and David Hilbert at Göttingen. He was a pacifist who objected to scientists colluding with the military establishment. He moonlighted as a journalist for the *Boston Herald*, and believed that automation would improve standards of living and put an end to economic underdevelopment. Many apocryphal tales are told of him, particularly about his absent-mindedness. The one I like best is about when he returned home one day to find his house empty. There was a girl waiting outside, so he went up to her and asked her what had happened. She explained that the family had moved house that day. When he thanked her for the information, she replied, 'That's why I stayed behind, Daddy!'

Cybernetics, Wiener's most important scientific legacy, is a unique synthesis of biology and mathematics that aims to understand how complex natural systems behave and evolve. Wiener's big idea was that by studying how life solves problems – such as, for instance, locomotion or information processing (things such as 'seeing' or 'feeling') – one could apply mathematics to mimic life and build automated engineering systems. Cybernetics envisaged a world in which we could decode nature and reproduce it by constructing a brave new civilisation, with self-regulating factories, therapies for every disease, robust economies, fair societies, and – yes – machines that thought.

As my studies in systems engineering progressed I became increasingly interested in computers and computer programming. This may sound obvious nowadays but when I started university in the early 1980s PCs were not widely available.[1] Undergraduate engineers had to program 'analogue computers', large calculating machines that looked like old-fashioned telephone switchboards and which used cathode ray tubes to perform calculations. Programming these ugly behemoths was a very cumbersome exercise, as they were prone to many errors that were difficult to trace or rectify. There was, of course, a 'digital computer' in our university, a Honeywell-built giant that took up most of the university's basement, but direct access to it was strictly controlled. To use it one had to book a time slot of a few minutes several days in advance.

The advent of PCs changed all that forever. They empowered experimentation and fast learning. One could now quickly test ideas about automation without having to look like a mad scientist surrounded by flashing lights and scores of wires. By the time I graduated I had acquired a lot of experience in computer programming, and decided to go into research. Applying Wiener's cybernetic concept to the field of computing, my PhD focused on automating the thought processes of medical doctors in intensive care units. This is an area of Artificial Intelligence that goes by the name 'expert systems'. What these systems do is study the way human experts process knowledge and take decisions, then use logic to encode human decision-making in a computer. There are many potential benefits from such systems: imagine expert decisions that must be taken in the absence of human experts, in remote and dangerous places such as on a battlefield or in interplanetary space. Nowadays, expert systems are routinely used in a wide spectrum of applications from medical diagnosis, finance and engineering, to video games and communications.

As I worked on my research, I began to have my first doubts about automating human intelligence. There was something about programming that did not quite capture the way minds seemed to function. Something was amiss. To understand why one must first look at the two fundamental and interlocking ideas upon which automation rests: *logic* and *feedback*. Although logic is not a trivial subject by any measure, it is the latter idea of *feedback* that is more important in the design and success of automatic systems. Feedback is perhaps the most prevalent mechanism in nature; it is how biological and ecological systems respond to their ever-changing environments. For instance, when bright light flashes in your eyes, you reflexively shut them in order to protect your retina. When the carbon dioxide in the atmosphere increases, ocean plankton consequently multiplies in greater numbers in order to absorb it. Markets use prices as feedback signals to determine levels in the production of goods and services. From amoebas to ecosystems, our world is a dense network of interconnected systems of ever-increasing complexity, all of which use feedback information to exist in dynamic equilibrium. Artificial cybernetic systems try to mimic the feedback mechanisms of nature. Take for example the simple system that helps you flush your toilet. A float 'senses' the water level and, connected to a valve, controls the water supply. The float is the measuring instrument that 'feeds back' to the valve information about the water level. Dynamic equilibrium is achieved when the water level is at a set limit.

When you flush the toilet this equilibrium is disturbed, so the system tries to regain it by opening up the valve and refilling the water tank.

An 'automated' mind would be considerably more sophisticated than a flushing toilet but, according to cybernetic theory, it would use the same basic principles of feedback and logic. It would mimic the human mind by using 'senses' to provide it with information about its environment, and logic to process this information and take informed decisions about its actions. For example, my medical expert system was fed with data about patients' vital signs, medical history and other test results and measurements; it then used its coded knowledge in a logical way in order to process the data and make decisions about treatment, just like a human doctor would. New data would change 'the mind' of the system, as its logic processed the changes and adapted its decisions. After all, isn't that what we humans do every moment of our waking lives? Don't we perceive our ever-changing world, and then use our knowledge and logic to act, change our minds and decide?

Well, that's how it seemed to me when I started my research but, by the time I'd finished it, I had formed a more nuanced opinion. Although my expert system worked very well – and got me a PhD – in no way could I claim that I had developed a truly 'intelligent' system. Perhaps it was smart, even fiendishly smart on certain occasions. But whenever it processed information about a patient, or suggested a therapy, it was not really 'aware' of its actions. It did not know what a 'patient' really meant, in the fullness of the meaning of a person with a mind, family, friends, aspirations, fears, discomforts, and everything else that it means to be human. What was missing from my expert systems was *consciousness*.

In the late 1980s, the 'c-word' was not something you uttered lightly in scientific, let alone engineering, circles. It conjured up images of hippies on LSD and books about shamans by Carlos Castaneda. At best, consciousness belonged to the mysterious – and, to the hardliner scientist, profoundly pointless – realms of psychology and philosophy. At worst, it was the surest way to be relegated to the nether world of psychic research on a one-way ticket.

And yet consciousness lay at the very heart of the problem of Artificial Intelligence. Creating automated systems that mimicked human experts and performed useful tasks was indeed possible. My research, and that of many others, demonstrated as much. But the 'holy grail' of machine intelligence, its quintessence, the ultimate game-changer, was to create a machine that *really* thought like a human. That meant one thing only: that the machine had to be *aware* of its thinking, that it somehow knew that it was 'it' that thought – just like I know that 'I' am writing these words. Otherwise the machine would be a zombie: it might appear to behave, or think, or write like a human, but would lack the subjective experience of its actions and thoughts; it would have no sense of 'self'; no 'inner existence'; its actions would be instinctive reflexes bereft of meaningful context or intent. Without consciousness, intelligent machines are senseless automata. Disillusioned with the lack of serious debate about consciousness in the field of Artificial Intelligence and cybernetics among my peers, I left academia to work in the private sector, and to continue my studies privately in that ultimate waste-of-time-and-effort area for most engineers and scientists: philosophy.

I often wish that philosophy were a core subject in undergraduate science or engineering curricula. Not only would it fertilise young minds with the richness of thought bequeathed by some of the most brilliant thinkers who ever lived, but it would spur new and innovative ways of approaching scientific questions. In my case, philosophy helped me realise the true magnitude of the problem in Artificial Intelligence.

For a machine to become conscious it would be necessary to code consciousness in a logical and consistent way; in other words to design a computer program endowed with self-awareness. The word used in computer engineering for logic programming is 'algorithm'. An algorithm is a sequence of finite logical steps that lead to the solving of a problem. So the central question in creating truly intelligent machines is whether consciousness is *algorithmic* or not. If it is algorithmic then it can be coded. But this question is a very difficult one to study. At the core of the problem resides our current, and very partial, understanding of consciousness as a biological phenomenon. Thankfully, after the late 1980s consciousness was purged of its pseudo-scientific reputation and became a respectable field for scientific research. We must thank Francis Crick for that, the English biologist who together with James Watson discovered DNA. In his 1994 book *The Astonishing Hypothesis*, Crick proposed that there had to be a neural mechanism[2] in the brain that was the cause of our sense of self. In effect, Crick suggested that consciousness is a purely biological phenomenon that could be identified and measured just like any other phenomenon in nature. If Crick's hypothesis is true then consciousness is indeed algorithmic, and therefore can be automated, just as Wiener envisaged nearly one hundred years ago. But is this hypothesis true? More importantly, is there a scientific, i.e. experimental, way to test[3] it?

Since the publication of Crick's seminal book, advances in medical scanning technologies have revealed many unknown details about the workings of the human brain. Neuroscience has advanced at an unprecedented rate. However, curiosity is not the only driver of this amazing evolution in our understanding of the brain. Declining demographics in Western societies are causing brain diseases to become increasingly prevalent. In Europe, brain-related diseases afflict more people than cancer, cardiovascular diseases and diabetes put together. One out of three people will suffer from such diseases, at least once in their lifetime, a daunting statistic that in Europe translates to 165 million people at a cost of

€800 billion per year.[4] Owing to the advances of neuroscience, the twenty-first century has been called the 'century of the brain' – and sometimes 'of the mind'. The deep-rooted belief that rigorous science, powerful computers and ever more accurate scanning instruments will ultimately 'crack the code' of the mind, has spurred governments into supporting even more intensive and systematic scientific research. The 'Human Brain Project' (HBP), one of the European Union's flagship ten-year projects, will be funded, initially, with a budget of €1.19 billion. The HBP brings together a host of scientific disciplines and talents from across Europe and the world in order to produce an accurate simulation of the human brain in a supercomputer. Decoding the human brain is the most significant scientific challenge of our times.

It is also a challenge like none before. The human brain is the most complex object in the known universe. It is made up of approximately one hundred billion cells called 'neurons', which connect to one another by means of nearly one hundred billion connections. Apart from being incredibly complex, the brain is also deeply mysterious: it 'thinks'. No one knows yet how. But the scientists in HBP believe that they can discover how the brain thinks by mapping it carefully. This is what science has done best since the Age of Enlightenment: it painstakingly and methodically catalogues every little aspect of the natural world, studies it, then connects the studied parts like dots in order to understand and explain the whole. Could this centuries-old scientific approach prove equally successful in the case of the human brain?

Although the scientists of HBP, as well as the politicians who fund them, seem to be convinced that their approach is infallible, they are up against a deep philosophical problem regarding the mind, which is known, rather prosaically, as 'the hard problem'. Australian philosopher David Chalmers has defined the *hard problem of consciousness* by distinguishing it from 'easier' problems that could be explained by examining brain functions: for example, memory, attention or language.

These 'easier' problems are by no means easy. The HBP project is going to keep itself very busy trying to solve them by applying the scientific method. But Chalmers made the point that there is a certain problem that cannot be explained by a purely materialistic view of the brain. This is the problem of subjective experience, sometimes called *qualia*.

Take, for instance, the 'redness' of red wine. The colour we call 'red' is an electromagnetic wave radiation with a wavelength between 620 and 740 nanometres. Although science can measure this wavelength with precision it has nothing to say about its 'redness', or why this particular wavelength appears to most of us to have a subjective quality we call 'red'. Chalmers argues that science can *never* be able to tell why we see light between 620 and 740 nanometres as 'red'.

Closely related to the hard problem of consciousness is the nature of the subjective experience of the self that wreaks havoc with the fundamental philosophical school to which science adheres. Science is based on empiricism, the notion that reality is what can be tested by experiment. Angels and fairies are beyond the scope of science because they cannot be experimentally verified (or falsified). For scientists angels and fairies are therefore 'unreal'. The problem with subjective experience is that it cannot be tested by an objective experiment. If I am the subject of an experiment to measure my conscious experiences then the only possible 'measurement' is my personal description of my inner feelings, a narrative that I make up by responding to a question. Scientists do not like narratives because they smack of the qualitative anti-empiricism of social sciences and humanities.

Thus, the problem of consciousness is not only scientific and philosophical but anthropological as well. It is an example of the 'two cultures' problem identified by the English chemist and novelist C. P. Snow in his famous 1959 Rede Lecture.[5] Snow suggested that Western civilisation suffers from a deep intellectual dichotomy between the exact sciences and the humanities. Mutual incomprehension between these two disciplines explains why so many of the world's problems are so hard to solve. If David Chalmers and C. P. Snow are right, the Human Brain Project may end up in ten years' time as a synonym for the Tower of Babel: a fruitless, lavish and arrogant effort to build the unbuildable. And all because of the cultural incomprehension between the brilliant brain scientists and those equally brilliant colleagues of theirs down the university corridor, the philosophers.

I met David Chalmers on several occasions in Tucson, during the biannual world conferences on Consciousness Studies organised by the University of Arizona. David, looking more like the guitarist of a heavy metal band than a university professor, was one of the main organisers together with anaesthesiologist Stuart Hameroff. Participating in several of these conferences during the early and mid-2000s, I had the opportunity to meet and talk with many scientists and philosophers who shared my fascination with the brain. By then, my journey in Artificial Intelligence had been diverted from computer programming towards trying to understand how the brain actually functioned. It had taken me from rainy London to sunny Arizona, and many other places in between. My private studies in neurophysiology and neuroscience enforced my conviction that developing machine consciousness was feasible, provided we applied the findings of neuroscience to reimagine computing machines. In effect, I too belonged to the intellectual camp of the Human Brain Project. Subjectivity had to arise from objectivity, for there could be no other way. If we took a brain apart and put it back together, we ought to get consciousness.

Then, one evening, I had a strange epiphany. It was April 2006 and I was at the official dinner on the last day of that year's Towards a Science of Consciousness conference. The dinner was taking place at the Sonora Desert Museum, a few miles south of Tucson. The Arizona sun was setting majestically over the barren hills and the vast plains that stretched all the way to the Mexican border. Iconic saguaro cactuses populated the darkening desert, and seemed like the silent guardians of a well-kept secret. The tables were arranged outdoors and I sat with a group of scientists discussing some of the talks we had attended. Soon enough, the conversation shifted from the hard problem of consciousness to Commander Data of *Star Trek*. It turned out we were all huge fans of the television series. In fact, we were all big sci-fi fans and avid readers of the masters of the genre, writers such as Philip K. Dick and Isaac Asimov. And then it struck me: fictional stories, such as the tale of Commander Data, had played a pivotal role in influencing everyone's academic life. I was there, on the other side of the world, *because* of that fateful night in Athens when I first encountered a mechanical life form, Robby the Robot. And I was not the only one. Movies, novels and sci-fi television shows had inspired the rest around the table too. Indeed, stories with robots as heroes or villains had determined the direction of our scientific lives. Could there be a link between literary narratives and scientific research agendas? Could our obsession with consciousness and AI be the result of fictional stories we read when we were younger? If so, where did these narratives come from? And why were we so fond of them?

A curious thing about literary narratives is that they resemble a network of veins. They seem to connect to a central source, a mighty river of archetypal stories flowing in the mists of time. From that source new stories evolve by constantly bifurcating, exploring new characters and new twists in the original plot, new directions and eventualities. But, whatever direction they take, they never seem to lose their deep connection with a primal story. Take, for instance, the plot of *Forbidden Planet* and how it resembles Shakespeare's *The Tempest*. The distant planet Altair IV of the movie is similar to the remote island of the play. Dr Morbius the scientist is reminiscent of Prospero the magician. Robby the Robot is a reinvention of Ariel, the ethereal spirit that obligingly serves his master, Prospero. But whence had Shakespeare taken *his* inspiration?

At that curious moment, with the setting sun and the saguaro cactuses in the background, it occurred to me that a deeper and more profound connection between our humanness and our technologies had to exist, which could be traced in art and literature. Our technological quest for artificial simulacra forged in our own image ought to have been somehow hardwired into our cognitive make-up as our species evolved. Perhaps we seek to construct Artificial Intelligence out of some instinctive impulse, rather than the utilitarian need for it. Consider the ramifications of a conscious machine: one that thinks and feels like a human, an 'electronic brain' that dreams and ponders its own existence, falls in and out of love, writes sonnets under the moonlight, laughs when happy, cries when sad. What exactly is it *useful* for? What could be the point of spending billions of euros, and countless hours of precious researcher intellect, in order to arrive at an exact replica of oneself? Why not simply get a human friend to talk to? Or a human employee to do the job?

Artificial Intelligence is arguably the most puzzling technology ever aspired to, a seemingly irrational enterprise, for the simple reason that it aims to duplicate *us* with all our misgivings and imperfections. The English novelist Douglas Adams made that sensitive point with gusto when he penned Marvin, the depressed, hyper-intelligent android of *The Hitchhiker's Guide to the Galaxy*; Marvin was so intelligent that people did not know what to do with him. Conscious machines would be just like us: bored, feeling undervalued and unloved – with an IQ many times higher than all of us put together. So why make one? Why are we so fascinated by mechanical simulacra furnished with intelligence?

This book recounts my quest to understand if it is possible to build an artificial mind, how we should go about it, as well as why artificial minds are so important and fascinating. It explores the questions I have been asking myself throughout my life as an engineer, but also as a novelist and a science communicator. I have structured the book in three parts, broadly corresponding to three perspectives from which I will try to explore Artificial Intelligence: literature, philosophy and computer science.

Part I traces the origins of stories about robots to the advent of art among our Palaeolithic ancestors. But why do I have to go so far back in order to explore something that is still very much in the future? Because I believe that, if we really want to understand Artificial Intelligence, we must begin by asking two important questions. Firstly, how the modern mind evolved, and what is special about the minds of modern humans that make us who we are? Secondly, why do we have stories about mechanical beings possessing minds similar to our own? Where did these stories originate? Could there be a deeper connection between our cognitive system and the reasons we want to build Artificial Intelligence? I will report on my archaeological digs into ancient and modern stories about robots and androids, and examine how relevant they are to research agendas and scientific expectations.

Part II ventures into the philosophy of mind and neuroscience. It summarises the most important philosophical ideas that are central to the modern debate about the mind, from Plato and Aristotle to prominent contemporary mind philosophers such as David Chalmers, Daniel Dennett and John Searle. What insights can they offer to our quest for an artificial mind? And how might the philosophical foundations of Western civilisation influence and determine our ideas about the nature of our own minds as well as computers and computer intelligence? What are the latest findings of neuroscientists about consciousness? Could engineers use these findings in order to create a conscious machine?

Part III presents the fascinating history of computers, the technology that has changed our world. It begins with the formulation of logic by Aristotle, and goes on to show how his ideas were developed further in the nineteenth and early twentieth centuries, until they led to the birth of computer languages and Artificial Intelligence. I will explore how ancient automata evolved into mechanical calculating machines, to Babbage's Analytical Engine, and all the way to modern supercomputers and the Internet of things; and speculate about futuristic alternative computer architectures that mimic the neural networks of the brain. I will ask how close computers are to achieving self-awareness, and what might happen once they do.

This book aspires to incite a fresh look at Artificial Intelligence by bridging the 'two cultures' gap, and illustrating the interconnection between literary narratives, philosophy and technology in defining and addressing the two most important scientific questions of all time: whence our minds and can we recreate them? You may find these questions interesting, or be sufficiently curious, to want to follow me on this journey. But the importance of Artificial Intelligence goes beyond intellectual curiosity. Artificial Intelligence is already with us, whether we ponder the ethical questions of autonomous drones killing people in the mountains of Pakistan or protest against government agencies mining our personal data in cyberspace. Increasingly, we interact with machines while expecting them to 'know' what we want, 'understand' what we mean and 'talk' to us in our human language.

As Artificial Intelligence evolves further, it will become the driver of a new machine age that could usher our species to new economic, social and technological heights. Supercomputers endowed with intelligence will be able to accurately model and simulate almost every natural process. We will acquire the power to engineer virtually everything: from new drugs to predicting markets and solving the problems of economic scarcity, to terraforming planets. Artificial Intelligence could make us virtually omnipotent. As citizens of a free society, we have a duty to come to terms with this future, and to understand and debate its moral, legal, political and ethical ramifications today. Heated arguments about stem cells or genetics will pale in comparison to what

Artificial Intelligence will allow us to do in a few years' time. Artificial Intelligence will define and shape the twenty-first century. It will determine the future of humanity in the centuries beyond.

Or it may be the cause of our demise, for there is a darker scenario at play. Many in the field of AI are convinced that whenever more powerful computers become conscious they will take over the world, and exterminate us. This 'AI Singularity' moment seems to borrow pages from the scripts of the *Matrix* and *Terminator* sagas. Super-intelligent machines interconnecting and becoming infinitely intelligent; then self-aware; then turning against us and blowing us all up, or using us as batteries. Could this be the sad and violent fate of humanity? Meeting our end at the robotic hands of our own creations? Should we heed Mary Shelley's cautionary tale *about the animation of the inanimate*, and take appropriate action now – before it's too late? I will be addressing these questions towards the end of this book, as well as highlighting the way Artificial Intelligence might impact our politics and ethics long before it becomes self-aware. But now, we must set out on our journey towards the creation of an artificial mind. And what better point to begin than the time when our own mind was born . . . .

PART I

DREAMING OF ELECTRIC SHEEP

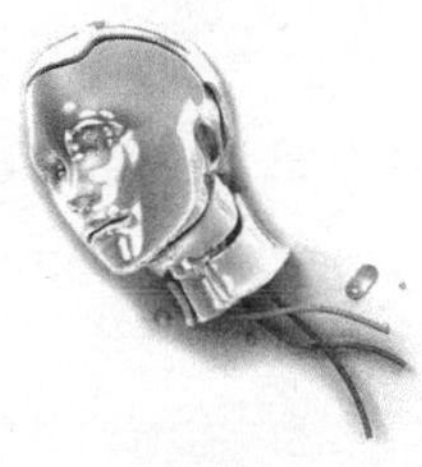

Did I request thee, Maker, from my clay
To mould Me man? Did I solicit thee
From darkness to promote me?

John Milton, *Paradise Lost* (X.743–5)

1
THE BIRTH OF THE MODERN MIND

In February 2013, the British Museum held one of the most remarkable exhibitions in its history. Under its esteemed roof the curators gathered the world's most ancient sculptures, drawings and portraits, loaned from the top prehistorical collections of Europe and Asia. It was the most comprehensive display of prehistoric art ever assembled. Among the objects on display, cocooned behind tempered glass and attracting awe-struck onlookers in droves, stood one of the most fascinating objects I have ever laid my eyes upon: the replica[1] of a statuette made of mammoth ivory, about 30 centimetres tall, of a creature with the head of a lion and the body of a man.

It was practically impossible to tear my eyes away from it. The alert forward gaze of the lion head, those pricked-up ears, that upward stance and athletic masculine body were laden with meaningful familiarity. The statuette touched something inside me with its unmistakable signature of kinship. Instantaneously, I felt that whoever had carved the original was as human as me. It was also a masterpiece – the work of an accomplished and highly skilful artist, for carving a mammoth tusk needs highly specialised dexterity and deep knowledge of the natural world. But what made the artefact even more special was that the 'Hohlenstein Stadel lion-man' – as the statuette is known – is the oldest object of art in the world. Discovered in the Stadel Cave in the Lone valley of south-western Germany, it has been dated to around 40,000 years ago.

By the time the lion-man was sculpted, humans had lived in Europe for five millennia. They had arrived from the Near East as our species made its grand exodus from Africa to colonise the world. Upon arrival in Europe those direct ancestors of ours met the Neanderthals, a different species of human with whom we shared a common ancestor. The Neanderthals were somewhat less advanced but they were also better adapted to the European Ice Age. The two human species co-existed for several thousand years, in what was a very rough place to live. The Gulf Stream had stopped circulating, which meant that there were no currents of warm seawater flowing from the equator to the Arctic. The North Atlantic was much colder than it is today and glaciers covered most of the British Isles and Scandinavia, as well as the better part of northern Europe all the way down to the Alps. Europe was cold and dry. Westerly winds did not bring rain or snow to the continent. The ground was frozen and temperatures dropped to minus 25°C in the winter with summer highs barely exceeding 10°C. But there was plentiful game to hunt. Tundra-steppe vegetation supported large herds of reindeer, horse and mammoth, as well as lions and bears. The first Europeans, having forfeited the balmy clime of their African motherland, had to readapt their lifestyles to the frostiness of their new home. They dressed in heavy clothing and endured rough living in extreme weather conditions, not unlike today's Inuit of Alaska or the Sami of Lapland. The going was far from easy. And yet they somehow found the time, the need and the exhilaration to express their minds in art.

The emergence of art in Europe 40,000 years ago is arguably the most fascinating, and significant, occurrence in the evolutionary history of our species. Why did we begin to create art? Most importantly, why hadn't we done so before? Our species existed for nearly 360,000 years before the time of the first Europeans,[2] and yet we had created nothing artistic.[3] For aeons we pursued a mundane life not dissimilar to that of other human species that preceded us, living and dying without fanfare, copiously reproducing the single artefact we had inherited from those who preceded us: the hand axe. And then, all of a sudden, 40,000 years ago, everything changes with a bang: we begin to paint cave walls with murals of stunning beauty and narrative complexity, to carve sumptuous figurines of women, to play flutes and dance, to adorn our bodies with beads and colours, to develop highly specialised new weapons and hunting techniques; to bury our dead with sophisticated rituals as if the dead continued to live beyond the grave; to imagine chimeras, half lion and half human. The emergence of art must represent a quantum evolutionary leap in our cognitive system. Take, for example, the chimpanzees, our closest living relatives today. We share nearly 98.8 per cent of our genetic material with them. And yet, however fascinating, likeable and intelligent chimpanzees are, they have never developed art.[4] Indeed, no other species on our planet has developed art except *Homo sapiens sapiens* – us. We are truly unique. If we want to understand how our minds became so unique – and why we seek to construct artificial minds – we must first seek whatever turned us into artists. To begin our quest we must climb down our genealogical tree by some six million years, to arrive at the common ancestor of human and chimpanzee, and begin our story from there.

The drama of our past

The common ancestor we share with the chimpanzees must have included several curious individuals within its ranks. They separated from their kin to explore better feeding grounds, and wandered about the ever-changing environment of Africa. Around 1.5 million years later, they evolved into several new species we call australopithecines.[5] The Earth's climate had changed considerably by then; the long grass of the savannah had replaced much of the tropical forest that once covered Africa. If we could pay a visit to those hairy distant grandfathers and grandmothers of ours we would be hard pushed to distinguish them from the other apes that roamed about the place at that same time. But they were indeed different, mutants set on a destiny that would one day separate them from the rest of the animal kingdom. A few would occasionally stand on their back feet to peek over the grass for lions, or food. With time, this advantageous habit of the few was inherited by their offspring, who were born with the gift of bipedalism. They could now run faster in the savannah, spot enemies more quickly and survive for longer. Their numbers multiplied with each generation. They evolved even more. They became less ape and more a type of being that could use tools and strategy to hunt, collaborate in teams, and increase the probability of survival by learning to adapt. Around two million years ago, the first member of our 'homo' lineage appears: *Homo habilis*. He inherits the knowledge of creating basic stone tools for chopping, scraping and pounding, and perfects it. Stone tools were used by earlier species too. There are relics from around two and three million years ago which are often difficult to distinguish from naturally occurring rocks. They belong to what archaeologists call the 'Omo Industrial complex', from the Omo area in Ethiopia where most of these archaic stone tools were found. With the appearance of *H. habilis* between two and 1.5 million years ago, stone tools become clearly identifiable as artefacts consisting of flakes removed from quartz, basalt or obsidian.[6] Hand axes appear around 1.4 million years ago, and they become the pinnacle of utilitarian design in terms of supporting human existence. The hand axe remains to this day the most successful technological innovation on Earth, if one judges its merits according to how long it was used. There were no further innovations until the Middle Palaeolithic, around 200,000 years ago. Was this because we remained relatively stupid and unimaginative for several millennia? Let's try to answer that by examining the brains of our ancestors – or, should I say, their skulls.

At 800 cubic centimetres *H. habilis* had almost double the brain size of the last australopithecines.[7] His was truly a giant leap in human evolution. A more evolved *H. habilis* called *Homo erectus* was the first human to leave the African homeland

1.8 million years ago. This is when the Pleistocene epoch begins. The Earth's climate changes once again. Ice sheets begin to form in high altitudes. *H. erectus* seems to appear simultaneously in three parts of the world: East Africa, China and Java. His brain size now leaps to a whopping 1250 cc. Let's scrutinise this fellow more closely. The most spectacular, and complete, *H. erectus* skeleton found is that of an eight-year-old boy dating 1.5 million years, from Nariokotome in Kenya.[8] The skeleton provides evidence for a linear rate of child development that appears to be characteristic of early humans. This contrasts with the growth spurt of modern humans that occurs in puberty. Although *H. erectus* is considered 'human' he was considerably different from us. He still retained many of the characteristics of apes. 'Human' was still in the making.

It took another 1.1 million years for the ape inside us to melt away, at least for the most part. Around 400,000 years ago[9] archaic *Homo sapiens* appears in Asia and Africa. It is an ill-defined species. It seems that, as *H. erectus* spread across Europe and Asia, he diversified in several ways, at different times, and in various geographical locations. By now brain size has reached 1400 cc. One of the sub species of archaic *H. sapiens*, a species called *Homo heidelbergensis*, made Europe his home; fossils discovered in Atapuerca, Spain, have been dated to at least 780,000 years ago. From this species evolved the Neanderthals (*H. neanderthalensis*) who appear 220,000 years ago and survive in Europe until 40,000 years ago.[10] Brain size has now reached a plateau between 1200 and 1750 cc. Neanderthals are muscular and stout with strongly built bodies and short legs, all anatomic adaptations finely tuned for living in a glacial environment.

Around the time that the Neanderthals appeared some significant changes occur in tool-making. There is now more diversity in the tools, and hand axes become less prominent. New tools are made with the so-called 'Levallois method', which produces carefully shaped flakes and points of stone.

Neanderthals use the method to make weapons and hunt big game. Yet almost since the inception of the hand axe toolkits tend to involve the same essential ingredients. True, some are now more finely crafted, but all are still made of stone, or wood, just like before. There is no experimentation with other materials, such as bone or antler. It seems that for one and a half million years, and despite the impressive development in brain size, the 'mind' of these evolved humans somehow remains 'stuck'. Their intelligence seems to have been of a specialised type; it worked well in several dimensions such as social interactions, tool-making, hunting, but not across all these dimensions at once. They seem to have lacked 'general intelligence', the type of intelligence that connects the dots, innovates, discovers, questions – all those things that our modern mind does.

The earliest anatomically modern humans appear in Palestine and South Africa about 100,000 years ago. Their bodies are less robust; they have no brow ridges, more rounded skulls and smaller teeth than the Neanderthals. The size of their brains is now between 1200 and 1700 cc, slightly smaller than that of the Neanderthals. Almost upon appearance, these new humans start making bone artefacts, as excavations in southern Africa have revealed. They place parts of animals into human burials in the Near East. For several thousands of years, our direct ancestors co-exist with other humans such as the Neanderthals, as well as remnant populations of other archaic *Homo sapiens*. But this time, evolution has decreed that only one species of human will survive – and that it will be us. We begin to colonise the planet anew, in a repeat of the first exodus from Africa that had taken place several million years previously with *H. erectus*. By 60,000 years ago, we have arrived in South East Asia, built boats, crossed the southern seas and colonised Australia. We enter Europe 40,000 years ago.

Evidence for this immigration scenario comes from the limited genetic diversity among living humans today. Living Africans have a higher degree of genetic variation than people elsewhere in the world. This can only be explained by a severe, and relatively recent, 'bottleneck' in human evolution. The first people to leave Africa must have been very few in number. One estimate suggests there were no more than six breeding individuals for seventy years, which means a population size of around fifty first colonists.[11] From this small group of people, our species gradually replaced all existing early humans. Thirty thousand years later, *Homo sapiens sapiens* was the only surviving member of the *Homo* lineage. We had conquered the world.

New things happen in tool manufacturing with the appearance of our species. Archaeological findings in the Near East show that instead of flakes being produced by the Levallois method, long thick slivers of flint are now removed from stones that look like – and are – blades. An interesting innovation, but perhaps more interesting is that nothing else is invented for the next 60,000 years. We have now arrived in the so-called Upper Palaeolithic Age.[12] Then, suddenly, instead of stone tools new materials such as ivory and bone are used. Instead of continuing to live in caves, *Homo sapiens* constructs dwellings. Caves are mostly abandoned and repurposed: their walls become covered with naturalistic paintings. In small, nomadic settlements, people sit around campfires and carve animal and human figures from stone and ivory, while others sew clothes with bone needles. They wear beads and pendants. They decorate their bodies. They want to look good. The fact that the denizens of the Upper Palaeolithic spent so much of their productive time making themselves pretty points to beauty acquiring a high degree of social value. Aesthetics must have become part of everyday life, like hunting and celebrating. The ritualistic burial of the dead becomes increasingly sophisticated. These humans behave in ways that we can relate to today.

The Neanderthals who live close by attempt to mimic these ingenious and creative humans by recreating crude versions of tools and body decorations. But they soon fade away from existence, as do all other *Homo* species. Was this because we hunted them down? Or was it simply because the competition proved too much? Did they breed with us, or did they simply die off? These questions are still researched and debated. Whatever happened, by 40,000 years ago *H. sapiens sapiens* is alone on the world stage.

The rate of change accelerates. Europe is ablaze with the colour of cave art between 30,000 and 12,000 years ago, although most of the continent remains frozen under the last Ice Age. Rapid global warming returns around 10,000 years ago, and the agricultural revolution takes place. We still live in that 'long summer' that began ten millennia ago, in the scientifically termed 'Holocene period'. It took humans four million years to evolve the hand axe, another two million years to somewhat improve it. And then, within a mere 20,000 years, a geological blink of an eye, they created art, agriculture, the wheel, computers and spaceships. This unbounded creativity kicked in between 65,000 and 40,000 years ago in what scientists call 'the Middle/Upper Palaeolithic transition', sometimes referred to as 'the big bang' of the modern mind. But what exactly banged?

What banged?

In 1979, an American archaeologist named Thomas Wynn published an article in which he claimed that the modern mind was already in place 300,000 years ago.[13] He based his claim on the evidence that *H. erectus* and archaic *H. sapiens* made symmetrical axes. To explain his theory, he adopted the idea that the phases of mental development in children reflect the phases of cognitive evolution in our human ancestors, an idea referred to as 'ontogeny recapitulates phylogeny'. This is an important scientific idea that correlates behaviour to cognition: one can observe behaviour, such as symmetrical tool-making, and draw conclusions about cognitive architecture and function.

Viewed from this – essentially behavioural – perspective, something important seems to occur in our brains after the age of four: we acquire the belief that other people have thoughts, desires, intentions and feelings of their own. We thus acquire the agency for empathy, which is essential in forging human relationships. I guess I am not the only parent to have carried out a false-belief experiment with their child in order to test this hypothesis in developmental psychology. If you'd like to perform it too, here's what you should do. Show your three-year-old a box of crayons and ask him what is inside. He will most likely tell you 'crayons'. But you, being the scientifically inclined sort, will have replaced the crayons with something else, say, sweets. Show him the candy, and put it back inside the box. Then ask your three-year-old to tell you what his mummy would think was in the box, if she walked through the door. He will most probably tell you 'sweets' – at least that's what my son told me when I did the experiment with him at the age of three. But when I repeated the experiment a year later he told me, correctly, that his mummy would think that the box contained crayons. Why? Because she did not know what he and I knew. Because his mummy had a different mind. My son had acquired what psychologists call 'theory of mind'. Most humans[14] have it. In fact, most humans at the age of four start believing that not only other humans but animals and objects have minds too: dolls and toy soldiers are very much alive in a child's imagination. However, according to Wynn and others, our species took time to develop theory of mind. It is very possible that it was the acquisition of theory of mind that gave rise to the Upper Palaeolithic transition.

English psychologist Nicholas Humphrey[15] elaborated further on the evolutionary rationale for theory of mind. He argued that when individuals live within a group and enter into a diverse set of cooperative, competitive and mutualistic relationships, individuals with the ability to predict the behaviour of others will achieve the greatest reproductive success. He coined the term 'social intelligence' to describe the mental toolbox that is essential to maintain social cohesion.

Therefore, there is selective pressure to have the ability to 'read' other people's minds. Early humans were dependent on retaining harmonious social relationships within their group for their survival. This involved much manipulation of other people's emotions, fears and wants. Today, six million years after we parted ways with the chimpanzees, the instinctive need to belong to a group dominates our personal and social life. Social rejection hurts: exile is a terrible punishment; separation from family and friends a personal tragedy. Our high-level consciousness, or general intelligence, seems to have evolved as part of social intelligence.

But what did it mean to be a human before the advent of high-level consciousness? How did it 'feel' to *be H. habilis*? What level of consciousness did those early humans experience? What was it like to think, or make sense of the world, with only a specialised intelligence? Daniel Dennett, the American mind philosopher, has described the consciousness of early humans as being akin to a state of 'rolling consciousness with swift memory loss'.[16] According to him, for *H. habilis* consciousness would have been somewhat like the state we experience when driving a car while engaged in conversation with a passenger. We do not 'think' of driving at all. We are, however, conscious of being at the wheel and thus always ready to react in an emergency.

Reduced consciousness in early humans explains the puzzle of their lack of variation in tools across time and space. They did not make tools designed for specific purposes. They ignored bone, antler and ivory as raw materials. For millions of years, it was just hand axes made of stone and little else. Enter our species with the invention of art, the development of new hunting technologies and tools and an evolved theory of mind. Something profoundly radical must have occurred within our cognitive system. Intriguingly, all the archaeological evidence suggests that this mental transformation of monumental proportions happened within a relatively short period of time. The sudden emergence of modern behaviour in Europe around 40,000 years ago has led many scientists to question the gradual evolution of the human cognitive system. Something must have 'kicked in' that caused the 'big bang' of the modern mind: a spark, a fifth element. The most dominant candidate for this cognitive transformation is language.

The first piece of evidence to support the notion that language begot our highly evolved consciousness is genetic. In the late 1990s, a team of British scientists[17] isolated a gene that is crucially involved in the development of speech and language. Dubbed 'FOXP2', it also became known as the 'language gene'. Steven Pinker, the renowned MIT psychologist, has called the finding the smoking gun for the relationship between genes and language.[18] The gene exists in other mammals too, including chimpanzees, but seems to have undergone a significant mutation in humans around 200,000 years ago, a period that roughly coincides with the advent of *H. sapiens sapiens*. The discovery of FOXP2 provides some validation of the language theory proposed by Noam Chomsky, about the connection between genes and language. Chomsky observed that children are born with an innate knowledge about language and grammatical structure, which had to be biologically determined. According to his language theory, we are hardwired for language, a notion shared and supported by Steven Pinker and other neurolinguists.

We saw how human brains became increasingly larger as our biological lineage made its epic journey though time; from 750 to 1250 cc for earliest *H. erectus* to 1200 to1750 cc for Neanderthals. Brain size reached a plateau between 1.8 million and 500,000 years ago, and rapidly increased as archaic *H. sapiens* appeared. Archaeological findings show the early humans had all the hardware for language installed. These are the two areas in the left-brain hemisphere responsible for grammar (Brocca's) and comprehension (Wernicke's). Given the existing brain architecture in early humans, genetic variations such as the mutation of FOXP2[19] must have accelerated the evolution of general intelligence.

However, genes mutate all the time. If mutant genes triggered the evolution of language there had to be a compelling evolutionary reason for their selection, and propagation in the next generations. The reason was probably that they facilitated the social cohesion of human groups, which was of vital importance. Persons with mutated language genes made better social conservationists. They were the unstoppable chatters of prehistory, the naturally born public relations experts. They wooed better mates with their words, or even their poetry perhaps, and passed on their mutant, chatty genes to their offspring. Their numbers proliferated with every new generation until being uber-talkative became the norm. Once early humans started talking they literarily could not stop – and this led to our cognitive fluidity.

The language of those early humans was different from ours in several aspects. As Robin Dunbar[20] has suggested, the language of early humans was a social language, a way of grooming. They used it as a means to send and receive social information. It was a language solely given to social gossip. This should not come as a surprise to us. We, the humans of the twenty-first century, continue to use language mostly for social gossip. Knowing what our neighbour did yesterday or bought in the sales is arguably more interesting than nuclear physics or climate change for the vast majority of humankind. A quick search on any social network is enough to convince the most hardened sceptic that chat about celebrities far outweighs any other form of conversation on any other given subject.[21] We have inherited this love of gossip from our ancestors.

Social language must have evolved rapidly between 150,000 and 50,000 years ago into a general-purpose language that was now used to convey information about the non-social as well. General-purpose language has selective advantage because it introduces general questions about animal behaviour, hunting and tool-making. The dynamics of evolution kicked in and ushered our species into an ever-increasing awareness of our surrounding world, expressed in words previously used only for people. Individuals with a facility for general-purpose language could compete more successfully for mates and provide better care for their offspring. It was general-purpose language that begot general intelligence. This is a stunning realisation. It means that words came before painting, music, dance and sculpture, as well as science and religion. Indeed, it suggests that language created our world.

But the primal, social origins of language were never abandoned. When we talk about physical objects, we still tend to ascribe to them an intrinsic tendency towards motion, implying they possess minds, as if they were living, social beings. As linguist Leonard Talmy[22] has observed, sentences such as 'the book toppled off the shelf' and 'the ball sailed through the window' imply that these objects move under their own power, since they are the equivalent of sentences such as 'a man entered the room'. Our world remains populated by social entities, whether these be artefacts, trees, rivers, mountains, houses or the engine of our car that refuses to start. Who among us has not kicked a door to take revenge on wood? Who has not played with dolls or toy soldiers, and not believed that they had minds too, that they were truly alive?

So let us recap what we have discovered so far. During the big bang of the modern mind humans acquired and developed general-purpose language that altered their consciousness, possibly because of a number of enabling genetic mutations. As a result, the close-knit groups of early humans expanded rapidly to embrace their wider environment, for we now had the words with which to describe everything. The world of animals and of things became filled with the mind. Humans thus became the creators of a symbolic universe imbued with meaning.

Art and the mind of objects

All three cognitive processes critical to making art – the mental conception of an image, intentional communication and the attribution of meaning – were present in the early human mind. However, thanks to the rapid evolution of language between 60,000 and 40,000 years ago, they began to function together, thereby creating a new cognitive process we call visual symbolism, or simply art.[23] Most prehistoric art is representational. In the Chauvet Cave in the Ardèche region of France, a cave discovered on 18 December 1994 and dating to 30,000 years ago, there are 300 remarkable naturalistic paintings of animals (rhinoceroses, lions, horses, reindeer, an owl). The paintings are on a par with better-known caves at Lascaux in France and Altamira in Spain. Representational art is not coincidental. Nowadays when we speak of art we usually mean non-utilitarian objects; owning these objects reflects wealth and social status. But this was not how our ancestors regarded the wonderful paintings on the walls of their caves. Art for the prehistoric people, and indeed for all subsequent generations until Western societies became secularised, was sacred and utilitarian. It served a useful purpose. It provided the symbolic canvas for making existence bearable for a species that had evolved to realise that its life was ephemeral and that death conquers all. The realisation of your inevitable death can only take place if you have a mind capable of self-awareness. In prehistoric art we discover the beginnings of religion and science, and importantly the cognitive roots of our hardwired belief that things can have minds, which also means that robots can ultimately become as intelligent as ourselves.

Our mind became modern when it perceived inanimate objects as social beings. Our language, social by origin, continues to frame our thinking in such a way that inanimate things have – by default – their own volition. Our cognitive make-up, mutated under evolutionary pressure towards general intelligence, compels us to instinctively regard representations of reality as reality itself. It is not hyperbole to claim that the 'world' did not really exist before we developed general intelligence. That it came into existence when we found words to describe it. In a curious alignment with the Book of Genesis, cognitive archaeology agrees that words[24] created the universe; and that the first thing modern humans did was to name the objects of that newly born universe.

Consequently, every time we think or speak about something we virtually create, again and again, the universe we live in. For it can only be a mental projection of our cognitive system, a linguistic interpretation of 'reality' (whatever that is) enmeshed in the haphazard complexities of human evolution. Our consciousness lives in a simulation of its own making, where we – or, rather, our brains – are the simulators, where the inner and the outer are indiscriminate. That is why prehistoric art resonates so much with us today: because it is full of 'spirit'. Because we, regardless of whether we claim to be religious or agnostic or atheist, know full well what spirit 'is': we feel it inside us and around us; it is a part of us, for we cannot escape the fact that we are *H. sapiens sapiens*.

Let us now see what interesting conclusions we can draw from our brief sojourn into the distant past that are also relevant to Artificial Intelligence. Firstly, and most importantly perhaps, is that general-purpose language predated, and begot, general intelligence. Language was what caused the genesis of the modern mind. The repercussions of this finding are enormous, and I will explore them in more detail in the final part of the book. Just consider, for now, the importance of language. It is not only a means of communication, but also the way that the world is represented in our consciousness.

An artificial mind may have other ways of representing the world. However, since ultimately we will be the creators of this artificial mind, we will aim to furnish it with representations familiar to us, for otherwise we will not be able to communicate with it, or comprehend it. A central research goal in Artificial Intelligence has always been to find a solution whereby an artificial system can communicate in natural language, i.e. in a general-purpose language. This has turned out to be a major problem for programmers and system designers. Language is very difficult to code for, and our journey into the Palaeolithic has shown why: language evolved in a haphazard way as a means of enhancing social cohesion within small hunter-gatherer groups. The main purpose of language was, and still is, gossip.

Second finding: by evolving general-purpose language we became inexorably dualistic: we started to perceive the world as a combination of the seen and the unseen, a mixture of what one felt through the senses and what one 'saw' with the imagination. A direct result of our dualistic thinking was that, according to our perceptions of them, inanimate objects acquired minds. Our theoretical debates (well, theoretical for the time being) about androids, and whether they should have equal rights with humans, stem from this dualistic way of thinking. How could they not? The sculptures and paintings of the Palaeolithic were considered as alive as other people, animals, trees, rocks, or indeed natural phenomena such as the bright, terrifying lightning that cut across a cloudy sky. As the modern mind emerged everything possessed a spirit in a landscape filled with social meaning. The invention of art is a manifestation of this uniquely human worldview.

But, ultimately, art did more than simply help our ancestors come to terms with their new, general-purpose, dualistic minds: it provided a means for expressing narrative, for telling stories and recording knowledge, for expressing religion and, after many thousands of years, for inventing science. Narratives are at the core of what we do, they codify what we believe, and guide how we think of ourselves. When we seek to create androids and robots with an artificial soul, age-old narratives about non-human beings continue to motivate us, and condition our expectations and goals. We need to understand these narratives, how they result from our cognitive systems, and how time has transformed them from stories about prehistoric chimeras into novels and films starring futuristic cyborgs. But first, let's return to the lion-man and listen to his story . . . .

2

LIFE IN THE BUSH OF GHOSTS

I am borrowing the title for this chapter from a wonderful novel written in 1954 by the Nigerian author Amos Tutuola. In the novel, a seven-year-old boy flees his village in western Africa after it has been destroyed by slave traders, and enters a forbidden place populated by supernatural beings. There he lives amongst the spirits for twenty-four years, gets married twice and frequently transforms into an animal. The novel inspired Brian Eno and David Byrne to write one of the most iconic music albums of all time, *My Life in the Bush of Ghosts*. It is also a novel that is very relevant to our quest to understand how our minds interpret the world. Ghosts, spirits, the metamorphosis of humans into animals are cultural universals. Tutuola's novel echoes the roaring primal river of stories created by the first modern humans. It was a river in which the visible and the invisible formed an uninterrupted continuum, where everything had a soul, a mind, and intelligence.

The lion-man of Stadel Cave is a relic of that ancient river of stories. It speaks of a lost myth of the Upper Palaeolithic, according to which this half human, half lion creature was a hero, a demon or a god. Perhaps the lion-man was a seven-year-old boy from a village who transformed into a lion. This is perhaps the reason why we relate so strongly to this ivory statuette from 40,000 years ago. It tells a familiar story. Like many Western children, I was raised with Aesop's Fables, where the resolute tortoise beats the smug hare in a race; the cunning fox steals the cheese from the hapless crow; and the carefree grasshopper learns a hard lesson about life from the diligent ant. Stories about animals are common not only in the West but in every culture on our planet. In them, animals not only 'talk' but are attributed with every other aspect of our humanity as well. Anthropomorphising animals appears to be instinctive. Give a twenty-first-century child a puppy and she will start talking to it as if the puppy had a mind like her own. Animals are often attributed human minds by adults too; if you are the owner of a dog or a cat you will know exactly what I mean. A tiny cognitive step separates animals with minds (think of a mouse called Mickey) and inanimate objects or machines with minds (think of McQueen, the hero of the movie *Cars*, or your car). The anthropomorphising process appears to have remained unaltered since the dawn of the modern mind. The lion-man, as well as the plethora of other Upper Palaeolithic statues that depict animals or chimeras, are 'alive', 'feeling' and 'thinking', just like us humans. They have minds. But why is this so? What evolutionary advantage does anthropomorphising confer?

Anthropology provides us with plenty of evidence for a possible answer. Again, the social dimension of our existence comes into play. Modern hunter-gatherers think of their natural world as a social network in which everything, living or not, is related. When the Inuit of the Canadian Arctic kill a polar bear they treat it as if it were another hunter. Often the bear is considered to be an ancestor being. For the Aborigines of Australia their landscape is full of social meaning, and they navigate through it by ascribing stories to its landmarks. Wells are supposed to have been dug by ancestor beings who used the trees as digging sticks. As anthropologist Tim Ingold[1] writes: 'For them there are not two worlds of persons (society) and things (nature) but just one world – one environment – saturated with personal powers and embracing both human beings, the animals and the plants on which they depend, and the landscape in which they live and move.' By extrapolating from today's hunter-gatherers to our ancestors of 40,000 years ago, we can see that the painted caves of the Upper Palaeolithic represent landscapes full of symbolic meanings, where the social and the natural worlds fuse into one. The forests, the tundra, the mountains, the rivers, the animals, the spirits are all denizens of a continuum in which humans are included and embedded deep within its narrative fabric. Animals move because they 'think' of moving. The falling of rain, the roaring of thunder 'speak'. Everything is alive and possesses intentions, thoughts and feelings, sometimes benign and other times not. The conflation and confusion of functions, aims and criteria seems like the normal, original condition of mankind.[2] It is also the basis of totemic thinking.

To understand the vital significance of totemic thinking for the survival of our forefathers and foremothers, let us revisit Ice Age Europe circa 17,000 years ago. The continent is at its coldest. Never before have humans lived in an environment harsher than the one that these ancient hunter-gatherers endure. This period is called 'the Magdalenian'[3] and lasts until around 10,000 years ago when the last Ice Age ends, and the agricultural revolution begins. Equipped with a general-purpose language and general intelligence, our species applies their evolved minds to surviving the long and ruthless winter of the Ice Age. They innovate. Archaeologists have unearthed evidence of a major shift in hunting techniques from that period. Elaborate tactics, weapons, logistics and strategies are developed and employed. Fishing spears, hooks and nets become increasingly common. The spear thrower is invented: a wood or bone rod with a hook on one end is fitted at the base of a spear, helping the bearer to throw the spear further away and with improved accuracy. But the real revolution lies in their tactics, in the coordinated group hunting techniques for the killing of large herd animals, especially in the river valleys of Western Europe and the plains of central and Eastern Europe. Until then, hunting had been undertaken either by individuals or small groups. But although the game remains the same – reindeer, red deer, bison and horse – these animals are now slaughtered en masse. The efficiency of the new hunting tactics is overwhelming. It is perhaps the first recorded instance of the devastating impact an intelligent species can have on its environment. At least fifty genera of large animals (mostly mammals) become extinct during this period because of overhunting.

The advantages of anthropomorphising when hunting become clearly, if not dauntingly, apparent. Modern humans, by imagining animals as possessing thoughts, could predict animal behaviour better. Hunters could foresee where the herd would feed, or in what direction it would move, and strategise accordingly. Their modern way of thinking, equipped with an advanced theory of mind, reaped clear utilitarian benefits from anthropomorphising animals. Group-hunting strategies were possible because of totemic thinking. The connection between survival and imagining non-human minds was forged forevermore.

But totemic thinking did not end with the anthropomorphism of animals. Totemism embeds humans within the natural world, and traces their descent from non-human species – ancestral beings created by the human imagination. The origin of these beings is always the unseen. Our minds instinctively imagine the invisible as writhing with dangerous life forms. When one walks alone in the dark, one's mind compulsively produces images of invisible beings lurking in the gloom. It is almost impossible not to. There seems to be an evolutionary explanation for this. Because of the way our eyes have evolved, we cannot distinguish shapes or movements very well at dusk. Things get very blurry. At nightfall we are virtually blind. But not so our prime enemies, the big cats. It is possible that over many millennia our ape and australopithecine ancestors were ambushed in the dark by mastodons, lions and leopards. Out of the dark came unexpected death. Inherited fear of darkness was articulated by the modern mind through general-purpose language. Abstract, fearsome darkness became populated with anthropomorphic demons and spirits. Once we were able to imagine the invisible our minds went wild with imagining. Creatures of our imagination populated the stories that older generations passed on to the young. Imaginary creatures became protagonists in depictions on the walls of caves.

Many archaeologists believe that the painted prehistoric caves were sites for the practice of magical ceremonies. The few findings of human debris suggest that no one lived in them on a permanent basis. Engravings, tucked away in narrow or low niches, represent individual devotions. Footprints of adults, adolescents and children imply that dances were performed inside the painted caves, possibly with the use of hallucinogenic drugs to induce ecstatic states of mind.[4] Perhaps those rituals were somewhat like a Palaeolithic movie theatre and church rolled into one: a shaman, flaming torch in hand, leads the procession into the cave's mystical innards, stopping under a mural of lions chasing horses, and recites a story about hero-hunters transforming into animals, or supernatural beings. Admiration was mixed with fear; and thus the two essential ingredients of a captivating yarn were invented. Our brains were ready for them.

Our storytelling brain

The neurological basis of storytelling was discovered by the celebrated American neuroscientist Michael Gazzaniga[5] while he and his team were working with split-brain patients. Patients such as these usually suffer from extreme cases of epilepsy that can be treated only by surgically severing the corpus callosum, the part of the brain that connects the right and left hemispheres. The result of such an operation is that the patient stops having seizures, but the connection between his hemispheres is lost. His right hemisphere (the non-speaking one) stops communicating with his left hemisphere (the speaking one). It is as if the patient now has two separate brains cohabiting the same cranium.

Gazzaniga experimented by asking the right hemisphere of his patients' brains to perform a task, for example to move the left hand, by providing the instructions within the visual field accessible only to the right brain. However, when he asked the left hemisphere the reason why the hand had moved, this hemisphere gave a coherent explanation that was, of course, confabulated. What the left hemisphere was doing was filling the gaps in the patient's memory with plausible inventions in order to explain what had happened. Narrative continuity had to be preserved. The hand moved and therefore there *had* to be an explanation. Of course, the left hemisphere had no idea that the right hemisphere had given the order for the hand to move. But the left hemisphere had to *invent* a reason. So the left hemisphere *created a story*.

The part of the brain's anatomy that is responsible for storytelling (fictitious or otherwise) was thus identified and named 'the interpreter'. Not surprisingly, the interpreter resides in the left hemisphere, where the brain areas for language also reside. It organises our memories into plausible stories. It acts like a writer collecting disparate pieces of information and patching everything together by filling the gaps with his imagination.

However, you do not have to be a split-brain patient for the interpreter in your brain to confabulate. It's what we all do all the time. Our memories are not precise recording instruments. Our brain is not like the hard drive of a video camera. Every time we describe a past event, our brain recalls a few facts and automatically fills the gaps with whatever can be used to preserve the coherence of our narrative. We are not compulsive liars, just natural storytellers. Which explains why several witnesses of the same event always give different accounts or testimonies.

We can only postulate why narratives were wired into our brains during the Upper Palaeolithic. Perhaps the telling of stories helped us to prepare psychologically for life's eventualities. Perhaps stories are like the holodeck of the starship *Enterprise*: simulated environments constructed by the brain in order to train our reactions and feelings for what may come. They may have served other purposes too. Narratives express our autobiographies. If someone asks us who we are, we usually respond by telling a story that explains where we were born, who our parents were, where we went to school, etc. It is not difficult to imagine the evolutionary advantages of people who were able to create, and communicate, coherent autobiographies at the beginning of the Upper Palaeolithic. With increased self-awareness, they would easily become the natural leaders in their group, thus having preferential access to the best mates. Whatever the reason, the result is who we are today: we, their descendants, equipped with a brain that automatically codes our knowledge, our experiences, our relationships and everything around us and inside us into ever-developing stories.

Imagining machines with minds

Let us summarise what archaeology and neuroscience have discovered about how the modern humans became storytellers. As our species evolved over millions of years we arrived at a point, probably between 150,000 to 100,000 years ago, when evolutionary selection favoured certain mutations that gave rise to the facility for general-purpose language. General-purpose language gave rise to theory of mind, with which we could predict the emotions and thoughts of our kin. Theory of mind was then projected on to animals and inanimate objects; and the world was anthropomorphised. Animals acquired thoughts and emotions. Objects were endowed with souls. Anthropomorphising thus led to totemism, the belief in unseen, ancestral spirits and in the dead living in the hereafter, thus sowing the seeds for religion. There were now two worlds and two planes of existence: the natural world of the senses and the imagined world of the spirits. We became dualists.

This dualistic way of thinking is wired into our modern mind. It is an inextricable aspect of our cognitive system, including that of the most ardent materialists among us. The rationalisation of dualism has defined theology and philosophy, as well as much of science, throughout the ages. As I will show later, in the case of contemporary consciousness studies, information technology and Artificial Intelligence, dualism is an especially dominant school of thought. Just think of the modern computer, and how we separate 'hardware' (the material part) from 'software' (the immaterial pattern, or form); or the many articles you may have read about 'downloading consciousness' in a computer.

Finally, we saw how our mutated Palaeolithic brain engaged 'the interpreter', a specific part of its anatomy that produced the stories, narratives and myths which travel through oral traditions down the ages. These stories are not mere entertainment but powerful drivers of our actions and thoughts. They code ideas, information, fears, anxieties and hopes that originate from the deepest past of our evolutionary history. They define our humanness. Our memory, individually and collectively, is one big narrative. Ultimately everything is reduced to literature. Consider our scientific investigations: despite their often inscrutable terminology, they all come together in the end to weave a story about how the cosmos was created, and how we came to be who we are. With their discoveries and their doubts, physics, chemistry, biology, economics, sociology, and everything else studied in the universities and labs of the twenty-first century feed this ever-expanding narrative of modern science – our way of making sense of the world and ourselves. And yet this world narrative that attempts to explain everything first began to be spun around campfires and inside caves a very long, long time ago.

So here is my hypothesis, based on what we can glean from the big bang of the modern mind: our contemporary stories about intelligent machines echo the adventures of half-human creatures carved on the tusks of mammoths and on the cave walls of Ice Age Europe. Robots and androids look, behave and supposedly feel like us, because we cannot help but anthropomorphise inanimate objects; because storytelling is hardwired in our minds. We seek to create intelligent machines because we are driven by stories about spirits incarnating into our creations. To illustrate how Palaeolithic narratives transformed into contemporary ideas of Artificial Intelligence and cyborgs, we need to take a quick historical tour of how artificial life became part of the world narrative, and how this narrative gave birth to machines that mimic life and humans. And since we are talking about stories we must begin with understanding the role of metaphor, for without metaphor no story is worth telling.

3
THE MECHANICAL TURK

Consider the following sentences:

> The heart is a mechanical pump.
>
> Atoms are miniature solar systems.
>
> DNA is life's library and genes are its books.
>
> Newton's equations of gravity are beautiful.

What these sentences have in common is that they are all metaphors. They connect two different things by means of an analogy. The heart *is not really* a mechanical pump, but *like* a mechanical pump in that it pumps blood around the body. It is next to impossible to think of anything without engaging with analogy and metaphor, for this is how our brains function.[1] From a cognitive perspective, metaphors are the linguistic manifestation of our prehistoric tendency to anthropomorphise. As we saw, because we possess theory of mind, we imagine animals as having minds too, and consequently think that animals think 'like' us. We project the same concept on to inanimate objects as well. We say the Moon 'rises' and the Sun 'sets'; that a book 'falls' to the floor. Metaphor is inextricably embedded in grammar, syntax and vocabulary.

But metaphors do not belong exclusively to the realm of casual talk, or literature. Science would have been impossible without them. It is only by analogy to old knowledge that new knowledge is created. Students of science and engineering are taught using metaphors. Metaphors in science are like the steps of a mental ladder that our species climbs as we distance ourselves from ignorance (which is a metaphor, too). The philosopher Thomas Kuhn[2] suggested that the role of metaphor in science stretches far beyond that of a device for teaching, and lies at the heart of how theories about the world are formulated. According to him, science is a war of competing metaphors. Each age uses its own metaphors to explain and describe natural phenomena. When the use of a specific metaphor ceases and a new metaphor takes its place, we have a 'paradigm shift' – as Kuhn called it – in the way science explains the world. During this never-ending process, there is a constant dialogue between scientific metaphors and technology, where the one informs the other. Sometimes a scientific metaphor facilitates, or impedes, the advent of a specific technology; often an emerging technology breeds the next scientific metaphor. The brain is perhaps the most profound example of how scientific metaphors modulate and transform through time in conversation with technology.

There have been at least six major paradigm shifts involving the brain since the time of the ancient Greeks. These shifts concern not only the brain but the whole body and, by extension, life. The mind, the brain, life – all three are intricately implicated. If we ever managed to produce artificial life then it should be straightforward to evolve artificial intelligence, too. When we think of intelligent machines most of us think of artificial beings, of robots, or androids, that speak, move and possibly think and feel like humans. In Western civilisation the discourse about the mind is often identical to the discourse about life, and this intimate connection will become more transparent as we examine the historical succession of philosophical and scientific constructs about what humans are 'like'.

The first metaphor for life that we know of is mud. In both the Jewish and Greek creation myths humans are manufactured from mud. In the second book of Genesis, Yahweh fashions Adam out of mud, then breathes life into him. In Greek legend, Prometheus shapes the first man out of mud and Athena breathes life into the clay figure. Scholars have traced the striking similarities between the two myths in previous civilisations in Mesopotamia. The metaphor of people made of mud seems consequential for agricultural societies whose life depended on farming and harvesting. Life sprouted from the ground. The dominant technology was agriculture. That first historical metaphor for life is still with us today; the word 'human'[3] is a relic of it. The mud metaphor changes drastically several centuries later with the invention of hydraulic and pneumatic engineering. These new inventions inspired the story of the first robot that, much like its human predecessors, was also fashioned by a god. Let's now see how this robot came to be, and what innovations took place in order to inspire it.

The defender of Crete

In modern Athens there is a little-visited monument at the eastern side of the Roman forum on the foothills of the Acropolis. It is called the 'Tower of Winds' and it was built in the first century BC, when the city was ruled by the Roman Republic. Its octagonal shape, as well as the reliefs depicting the eight principal winds, give away little as to its original use. It is in fact a meteorological station, the first of its kind to encompass a very sophisticated clock. The clock was powered by water and was constructed by Andronicus of Cyrrhus, who used the blueprints of Archimedes (287–212 BC), the original inventor of the water clock. An ingenious arrangement that included water tanks in cascade and outflow nozzles regulated by a calibrated disk ensured that the clock showed the right time on any given day in a year. Not only was it a great improvement on sundials, which were useless under a cloudy sky, but it also provided the basis for time-based observations of the weather. For the first time it was possible to collect and categorise detailed weather data over many years, and draw conclusions and predictions. The Tower of the Winds was a data-driven scientific lab of a sort that many contemporary weather scientists would recognise.

Andronicus and Archimedes were two of many stellar engineers of the Hellenistic period, the era that follows the conquests of Alexander the Great and the 'export' of classical Greek civilisation to the Near East and Egypt. During this period – which lasts until the total conquest of the Greek world by the Romans during the reign of Octavian[4] – a creative explosion takes place in engineering, mathematics and medicine. Inventions and ideas from Alexandria and Antioch are transplanted into Rome, and inform European civilisation ever after. The steam engine of Hero of Alexandria, the astrolabe of Hipparchus, the mathematics of Euclid are all examples of this creative outburst. Hydraulics and pneumatic systems are discovered. As a consequence, the use of water and steam to cause the movement of inanimate objects through clever engineering creates a new paradigm shift concerning the concept of life. From the third century BC life is increasingly described not as static mud animated by divine will, but in terms of dynamically moving fluids within a mechanical body, a metaphor that will dominate Western thought for the next sixteen centuries.

The reason why this metaphor became so powerful was because the invention of hydraulic and pneumatic engineering coincided with a new scientific understanding of the human body and medicine. Hippocrates (460–370 BC), the father of modern medicine, developed a comprehensive theory about the human body that laid the foundations for the rational interpretation of disease. His theory was based on the flow of four different fluids in the human body that he called 'humours': black bile, yellow bile, phlegm and blood. 'Humourism', as the theory is known, was picked up by Galen of Pergamus (Claudius Galenus) around the second century AD and was subsequently greatly expanded and improved, forming the main corpus of Western medicine until the advent of Enlightenment.[5] Galen was a polymath who believed, like his contemporaries, in the existence of the soul. But he was not content to view it as the explanation of everything. Influenced by the Empiricist school of philosophy, he conducted experiments in human anatomy and medicine that led him to theorise about the localisation of function in the human body. This theory, still part of modern medicine, posits that certain parts of the body are responsible for certain functions. However, Galen did correlate body parts to the soul: the rational soul resided in the brain, the spiritual one in the heart and the appetitive in the liver. During his lifetime, philosophers were obsessed with one of the most significant questions in philosophy, the so-called 'mind-body problem'. This problem, which would go on to fuel the discourse in mind philosophy ever after, concerned how a material object (the brain) could produce an immaterial result (the thoughts) and vice versa (i.e. how immaterial thoughts can give rise to bodily actions). Galen supported the Greek approach to the problem, that there was no distinction between the mental and the physical. Using the dominant metaphor of his age, he suggested that human beings were complex hydraulic automata, their actions controlled by the movement and mixing of fluids (humours) inside their bodies. The mind and the soul *were* movement of humours. Nerves were conduits that conveyed animal spirits (which were material fluids) between tissues dominated by the humours. Thanks to the engineering advances of the Hellenistic era, Galen's ideas found their practical demonstration in hydraulic automata. It was the first time in human history that a scientific idea about life could be illustrated using a machine. Using clever hydraulics, the engineers of the Hellenistic times were able to demonstrate how life works. It is therefore not a coincidence that, at the same time, we hear of the first stories about mechanical men whose movements – or 'life' – depended on the flow of liquids, i.e. artificial humours.

The most famous of these stories is recounted in the *Argonautica*, a novel written by Apollonius Rhodius in the third century BC. Here we meet Talos, an artificial giant made of bronze, who stomps along the coast of Crete in order to defend it from invaders. In order for Talos to move and have 'life', Hephaestus, the giant's creator, manufactures a vein that runs from the giant's neck to his ankle, and which is bound shut by one bronze nail. In that vein, Hephaestus pours a fluid called 'ichor', the blood of the immortal gods. According to the novel, Talos is defeated by the witch Medea, who tricks him into opening up his bronze nail and thus lets his life-giving ichor out. However, Talos was not the only robot manufactured by Hephaestus. The god of fire and craftsmanship had several good-looking robotic maidens helping him in his workshop, as well as a number of other mechanical automata, including a tripod that walked about and spoke. But what is interesting about Hephaestus and his mechanical creations is that scant mention is made of them in earlier times.[6] New, revolutionary developments in the thought of the ancient world during the Hellenistic period gave new prominence to this lame god whose glamorous wife cheated on him systematically.[7] Now mortals were inventing and constructing artificial life, too: the automata.

Many other automata were designed and constructed during the reign of Alexander's epigones and the early Roman Empire. The tradition persisted as the Roman world split in two halves in the fourth century AD. The Eastern Roman Empire – also known as Byzantium – came to rule over the cities of the old Hellenistic kingdoms and thus inherited the technology of automata. The emperors of Constantinople used automata in their royal court in order to impress ambassadors and foreign dignitaries. A historical account in the tenth century AD mentions how the emperor sat at 'the throne of Solomon',[8] inside a special reception hall filled with a plethora of automata. As foreign emissaries knelt before the imperial representative of God on Earth, the throne began to rise from the floor while metallic replicas of lions roared.

The fame of Byzantine ('Roman') automata spread throughout the eastern world, as far as India and Indochina. In *Lokapannatti*, a Buddhist story written between the eleventh and twelfth centuries AD, there is an episode in which Emperor Ashoka, the hero, attempts to get his hands on the Buddha's relics, which are protected by mechanical robot guardians that come from 'Rome'.[9] According to the tale, the robots were manufactured by an Indian engineer who, hoping to steal the secrets of the robots from the 'Romans', vowed on his deathbed to be reborn in Rome. There, in his next life, he manages to marry the daughter of the robots' inventor and thus acquire from his father-in-law the sought-after blueprints. He then smuggles the secrets to India by giving them to his son, who arrives in Pataliputra, the famed Indian capital, just as Buddha enters nirvana. The reigning King Ajatasatru orders several robots to be constructed on the basis of the stolen blueprints in order to protect the relics of Buddha forevermore; and stations them as eternal guards with twirling swords ready to strike. One hundred years later, Emperor Ashoka finds the still living son of the Indian engineer, and gets him to disarm the formidable robots.

Closer to Byzantium, and at about the same period, the Muslim inventor Al-Jazari is credited with constructing a number of hydraulic and pneumatic automata. His most celebrated construction was a boat with four automatic musicians that floated on a lake and entertained guests at royal parties. From Constantinople to Damascus, Baghdad to Pataliputra, hydraulic automata entertained the curious-minded by simulating life and demonstrating that scientists had cracked its mysteries: it was all about the movement of humours.

The metaphor of the human body as an intricate vessel in which fluids flowed and mixed would begin to wane as the blacksmiths and mechanical engineers of the Renaissance discovered improved alloys and new mechanical parts, and as the first mechanical clocks began to adorn the bell towers of Europe. A new paradigm, or a new metaphor for life, was born.

Doctor Mirabilis and the brazen head

In the twelfth century AD a tale circulated in Europe about an artificial head that spoke. It was said that Pope Silvester II, who died in the year 1003, had constructed a head which 'spake not unless spoken to, but then pronounced the truth, either in the affirmative or the negative'.[10] In today's terms, the Pope's automaton was capable of communicating one bit of information only, but it was always right! Two centuries later, the fable was retold using a different hero, the English scholar Roger Bacon (*c.* 1214–1292), a man ahead of his time. Bacon was a polymath who became known as 'Doctor Mirabilis', his interests embracing alchemy, theology, physics, astronomy and many other sciences as well. In 1589, at the heart of the Elizabethan period, the playwright Robert Greene picked up on the story, as well as the infamy of Roger Bacon, and wrote a play entitled *Friar Bacon and Friar Bungay*. It is a multi-plot play in which Prince Edward, son and heir of King Henry III, plans to seduce Margaret, the Fair Maid of Fressingfield, with the help of Friar Bacon.

In one of the subplots, Friar Bacon undertakes to protect England against invasion by constructing a wall of brass around the whole country. Echoing the ancient myth of Talos, Bacon invents an intelligent brazen head to assist him. He enlists the services of Friar Bungay, a fellow magician, and the two put together a head made of brass that is the mechanical replica of a natural head. Not knowing how to set the head in motion, the two friars raise a 'devil' in a nearby wood who discloses the secret – but refuses to specify the length of time for the process to take effect. Exhausted, the two friars fall asleep, leaving the care of the mute brazen head to their servant. While they are asleep the servant starts to tease the head and taunt it with questions. And then the brazen head comes alive. In a wonderful ironic moment, during which the main heroes are there but not there, the brazen head speaks to the hapless servant, three times only, saying: 'Time is', 'Time was' and 'Time is past'. Before the friars awake, the head shatters to pieces.

The apocryphal story of Friar Bacon and the brazen head is a fable of our time, too.[11] Neurologist Warren S. McCulloch, one of the fathers of Artificial Intelligence, famously quipped, 'We will be there when the brass head speaks'.[12] McCulloch and his colleague Walter Pitts were the architects of the theory of neural networks, the idea that we can replicate a human brain by copying the brain's architectural elements, the neurons. McCulloch promised that scientists, unlike the magicians in Greene's play, would be awake while the 'brazen head speaks'. They would be able to measure its intelligence, and perhaps achieve something before it says 'Time is past' – that is, before it's too late. Contemporary sci-fi writers have also picked up on the brazen head image. In Philip K. Dick's 1967 novel *The Zap Gun*, a talking bronze head gives advice so cluttered with classical Greek and Latin phrases that it is practically useless. And William Gibson ends his 1984 iconic novel *Neuromancer* with the head appearing in the form of a computer terminal, richly decorated with gems, its voice generated by 'a beautiful arrangement of gears and miniature organ pipes . . . a perverse things, because synth-voice chips cost next to nothing . . .'.

In this one sentence, Gibson juxtaposes our current silicon-based technology with the mechanical metaphor for the brain that is dominant from around the sixteenth century until the early nineteenth century. By then, automata have ceased to be hydraulic or pneumatic, and have become mechanical, powered by key-wound springs and gears. This new generation of mechanical automata is part of the dialogue between the dominant technology of the Enlightenment and its dominant paradigm for life. The latter was formulated – and promulgated – by the founder of modern philosophy, Descartes.

René Descartes (1596–1650) postulated that the bodies of people and animals are nothing more than complex machines – and that the bones, muscles and organs could be replaced with cogs, pistons and camshafts. Thomas Hobbes (1588–

1679) agreed with Descartes and suggested that ideas and associations result from minute mechanical motions in the head. In his book *L'Homme machine* (1748), the French physician Julien Offray de La Mettrie writes that the body is 'a machine that winds its own springs – the living image of perpetual motion . . . man is an assemblage of springs that activate reciprocally by one another'.

Taking Descartes' machine metaphor to a profitable conclusion, the author and inventor Wolfgang von Kempelen built a chess-playing automaton in 1769 that he called 'the Turk', and presented it to the Austrian Empress Maria Theresa. The Empress was impressed and the Turk became an instant worldwide sensation. For the next eighty-four years, the mechanical Turk made the rounds of the European courts and toured America, winning against most human opponents, including world leaders such as Napoleon Bonaparte and Benjamin Franklin. It was not until the 1820s that the Turk was exposed as an elaborated hoax; there was a man inside the machine, a skilled operator who positioned the pieces. It would take almost another two centuries for an automaton, a real one this time, to challenge a human at chess – and win.[13] In that same period, the world moved away from the mechanical metaphor of Descartes. A combination of new scientific discoveries, as well as intellectual reaction to the empiricism and rationalism of the Enlightenment, led to a curious revision of Descartes' philosophical doctrines and a new metaphor for life.

The spirit of life

Descartes not only instigated the mechanical metaphor for life but shaped the philosophical arguments for dualism. Putting aside the evidence of cognitive archaeology about the Palaeolithic big bang of the modern mind, Descartes was the thinker who articulated and rationalised our hardwired, instinctive conviction that there is an invisible world beyond the senses. Indeed, he put into words what we instinctively conjecture every moment of our waking lives. Descartes theorised that the world is made up of two substances that he called *res extensa* and *res cogitans*:[14] things that are of matter and things that are of the mind. Things of the mind – for example, thoughts and dreams – are made up of non-physical substances. The impact of his ideas on European thought was tremendous, and it is still felt today in contemporary philosophical debates about the nature of the mind, the possibility of Artificial Intelligence, and much more.

Cartesian dualism became the established philosophy in the West by the early eighteenth century. Moreover, new scientific discoveries seemed to confirm Descartes. There were indeed invisible forces acting on material objects, such as gravity and magnetism. Take, for instance, Newton's gravitational laws. They explained almost everything, from the smallest to the colossal, and from the earthly to the celestial, but they were based on the uncomfortable assumption that gravity was a mysterious force that acted at a distance. Something 'spooky' held the universe in place and the planets in orbit. Newton was aware of the problem, and passed the responsibility for discovering the nature of gravity on to future generations of scientists. However, it was the new discoveries in the fields of electricity and chemistry that combined with Cartesian dualism to formulate a new metaphor for life. Let's see how that happened, and how it has influenced how we think of artificial minds and bodies today.

Until the 1600s, electricity was mainly an intellectual curiosity. The phenomenon of causing sparks by, for instance, rubbing together amber and wool had been known about since ancient times. The first scientist who systematically studied electricity was the Englishman William Gilbert (1544–1603) who coined the word 'electric' from the Greek word *elektron*, which means 'amber'. One hundred years later, Benjamin Franklin (1705–1790) became so fascinated with electricity that he sold his possessions in order to fund his research. One of the great stories of modern science tells of how, in June 1752, Franklin attached a metal key to the bottom of a dampened kite string and flew the kite in a storm. He luckily survived the electric shock because he was standing under an umbrella to keep dry!

A few years later, in 1791, the Italian physician Luigi Galvani (1737–1798) conducted a series of experiments with frogs or, to be more precise, with frog's legs. The experiments showed that when electricity passed through a dead frog's legs they kicked. Galvani coined the term 'animal electricity'. He had discovered bioelectricity. Galvani's experiments were repeated by Alessandro Volta (1745–1827) – who later invented the electrical battery – and ignited the imagination of Europe. By the mid-nineteenth century, Emil du Bois-Reymond (1818–1896), a German physician, had developed the galvanometer to measure electric currents in animals, frogs and humans. Using this new instrument, du Bois-Raymond discovered that electricity flowed along the nerves of the body. Today, he is considered the father of electrophysiology. All these discoveries and inventions confirmed Cartesian dualism by demonstrating how an invisible – and until then 'mysterious' – force called electricity moved the bodies of living creatures. And since 'movement' was directly associated with 'life' it followed that electricity was some kind of 'life force'. In other words, to be alive you had to be electric.

Cartesian dualism was also confirmed by developments in chemistry. It was observed that there were two kinds of chemical reactions. Non-living substances underwent chemical transformations that were reversible. For instance, it was possible to break down an acid and put it back together again. But substances taken from living things were not reversible. The chemical transformations of these substances changed them permanently: for example, cooking a vegetable or meat. There was no way of going back to the original substance. Thus chemistry was divided into 'organic' and 'inorganic' branches. Chemists of the time theorised that in order for chemical reactions to be sustained in living organisms there had to be some invisible regulatory force that maintained function. Thus the idea of a 'spirit of life', or *élan vital*, was born. This idea was called 'vitalism'. Vitalism introduced a revised metaphor for life: that humans possessed an inner, mysterious life force – possibly of electrical nature – that sustained them. In many ways, the 'new' idea was a revision of the Hippocratic humourism. The four temperaments, or humours, of the ancients were now unified and replaced with a new term: 'the vital spark'.

German physician Franz Anton Mesmer (1734–1815) took vitalism to another level by suggesting that there was a natural energetic transfer between animated and inanimate objects, which he called 'animal magnetism'. 'Mesmerism', as his theory was called, attracted a huge following across Europe and in Victorian England in particular,[15] influencing the work of Charles Dickens and Mary Shelley. That is why in Shelley's novel *Frankenstein* the dead body becomes animated when lightning is channelled through it. The dead monster has to be electrified in order to come alive.

Vitalism and mesmerism were discredited by the late nineteenth and early twentieth centuries. Advances in molecular biology explained the phenomenon of homeostasis, i.e. how life is sustained when chemical reactions are irreversible. Electricity was shown to be the result of elementary particles called electrons moving along a conducting medium. As for mesmerism – no mysterious field has been discovered that causes telekinesis or telepathy, although many people still believe in both. Nevertheless, the legacy of vitalism has been impossible to eradicate. After all, vitalism was the logical synthesis of the philosophical traditions and scientific investigations that had their origins in the Hellenistic period and which persisted for two millennia, until the early twentieth century. No wonder that the core tenets of vitalism are still believed by many today. Take, for example, the so-called 'alternative therapies' that claim to modulate 'energy fields' in the human body.

Because of vitalism's discrediting, many scientists are sceptical, if not outright suspicious, about contemporary scientific theories of emergence in complex systems. These theories suggest that the behaviour of complex systems (e.g. the brain, the weather, the stock market) cannot be explained by knowing the behaviour of the individual parts. Knowing everything there is about water molecules, for instance, does not mean we can predict the weather. Emergence therefore suggests that there is 'something else' that acts on the individual parts of a complex system and which compels them into new types of collective behaviour. Take, for example, the brain. Each individual nerve cell sends out electrical pulses to other nerve cells (a process that has been very well described and understood), and all the nerve cells together evoke consciousness (a collective outcome that remains mostly mysterious). What connects individual actions to collective outcomes? This fundamental scientific question cannot be answered satisfactorily unless one assumes a 'fifth element', some kind of a 'force' that takes over when the individual elements come together and makes them behave as a whole in new, emergent ways. The problem with this analysis is that it resembles vitalism. Studying the behaviour of complex phenomena assumes the existence of a mysterious, hitherto undetected 'force' that lies beyond the constituent parts of the system. The emergentists' counterargument is that interactions between the constituent parts of the system are also important for the behaviour of the system as a whole. However, if interactions are so important, what is the nature of these interactions? And how can we test whether they exist or not? For many scientists, emergentist theories that spring from cybernetics and complexity theory do not seem falsifiable, and are therefore suspiciously non-scientific.

Suspicion about theories of emergence reflects the ideological divide between traditional scientific methods of reductionism versus alternative systemic, or holistic, methods. Reductionism is the very successfully applied idea in science whereby one tries to reduce a natural phenomenon to an irreducible level that can be then studied. For example, particle physics is the irreducible level for almost everything[16] in the universe. Systemic approaches ignore the individual parts of the systems and study the behaviour of the system as a whole. As I will argue later in this book, friction between these two opposing schools of scientific investigation creates tension and confusion in contemporary consciousness studies as well as in Artificial Intelligence. But let me return to vitalism and dualism one last time, because their most significant legacy lies not in the proliferation of websites promising magical cures using crystals. They still influence the way we think today of the mind as something separate from the body.

Until the time of Galen, the mind was considered a physical thing. There was no disconnection between mind and body. They were one and the same. After Descartes, the mind became disembodied. It dematerialised. Vitalism was the scientific manifestation of dualism. But although vitalism was discredited, dualism was not. The disambiguation of the mind persisted, and became even more pronounced in new metaphors for the brain, as the nineteenth century ushered in technologies that permitted messages to be transmitted from a distance.

The brain as a computer

In 1838, Sir William Fothergill Cooke and Charles Wheatstone established the first commercial telegraph in the world along the Great Western Railway by connecting Paddington station to West Drayton. A year earlier in America, Samuel Morse had independently developed and patented an electrical telegraph, that would ultimately become the world standard.[17]

Meanwhile, further developments in microscopy revealed the intricate, wire-like structure of brain cells, the neurons. Neurons are different from other cells in our body because they are connected to each other via a dense network of endings that sprout from each individual nerve cell and which are called 'dendrites'. Dendrites meet other dendrites from other neurons and form connections called 'axons'. The scientist who first systematically studied neurons and nerve cells was the German physician, physicist and philosopher Hermann von Helmholtz (1821–1894). Von Helmholtz is one of the giants of science, a genius of unequivocal foresight who made contributions in many fields including physiology, optics, thermodynamics and psychology. In 1849, he measured the speed at which a signal travels along a nerve fibre. He is credited with introducing a new metaphor for the brain: the brain as a telegraph.

The advent of computer technology in the 1940s replaced von Helmholtz's metaphor. Now there existed machines that performed logical processes. They took as input raw data and produced new knowledge. They were doing what we humans call 'thinking'. Indeed, at first computers were called 'electronic brains'. In turn, the human brain was also likened to a computer. Since the mid-twentieth century we have been living in the era of the metaphor of the brain as a computer. Meanwhile, the telegraph has evolved into email and instant messaging. Telecommunication and computer technologies have merged. Often, the Internet is discussed in terms of its being a global, or planetary, artificial 'brain' that is evolving with every new connection.[18] According to this new metaphor, the Internet resembles a human brain because individual parts (your iPhone, or your computer) connect with many other individual parts through a mesh of wireless and wired connections. Thus the Internet is 'like' a brain and the brain is 'like' the Internet. Perhaps one day, given enough connections, the Internet will become 'conscious' – or so the metaphor suggests.

This new metaphor has taken the dualistic disembodiment of mind to a whole new level. Because of the distinction in computer technology between hardware (the physical part, the integrated circuits of miniaturised electronics) and software (the execution code), the human brain is also regarded in this dualist way. There is a brain 'hardware' – the mushy grey stuff in your cranium made up of neurons – and a brain 'software', your non-physical, mind, thoughts, and dreams. Dualism is so deeply imbedded in the current metaphor of the brain as computer that renowned scientists and philosophers adopt it as a given when reflecting on how minds could be 'coded' in computers and thus achieve digital immortality.

Thinking through metaphor, feeling through narrative

As we have seen, because of the way our storytelling brain has evolved it is impossible to think about anything without using metaphor and analogy. Both are linguistic tools for discovering, debating and pushing the boundaries of knowledge. They have served us very well since the Upper Palaeolithic. Thanks to them, we developed our technological civilisation. And we have considered how, as humankind progressed from the agricultural revolution to the Greco-Roman world, Renaissance, Enlightenment and the modern times, our metaphors for life and the mind have evolved and mutated. First came mud, then water or humours, then mechanics, the electric current or spark of life, followed by the telegraph and now the computer. For each of these metaphors, people have imagined automata, artificial artefacts set in motion by technologies that support the metaphors. In Hellenistic Egypt, it was hydraulic engineering. In seventeenth-century France, it was mechanical gears and springs. In the twenty-first century, it is computer engineering.

Nevertheless, and because we always tend to think through metaphor, it is too easy to confuse the metaphorical with the actual. The brain is *not* a computer, not *really*. The Internet is *not* a brain. These are actual statements representing actual facts. So what is the brain? How does consciousness arise?

Where is our mind located? We can only seek answers to these questions by stripping away the metaphors and focusing on the materialistic essence of what computing actually is. Thankfully, as I will show later, this is not an impossible task.

Metaphor conditions our thoughts about artificial life and intelligence, but how about our feelings? How do we feel about robots which think, and perhaps look like us? The writer and journalist Pamela McCorduck has suggested[19] that there are two prevalent attitudes in Western societies with regards to Artificial Intelligence – a positive and a negative. She refers to the welcoming attitude as the 'Hellenic' point of view, as in coming out of ancient Greece. Quoting the Second Commandment of the Bible,[20] which was composed about the same time as the early Greek literature,[21] she refers to the attitude that finds intelligent machines wicked, or even blasphemous, as 'Hebraic'.

Without doubt Western attitudes[22] towards Artificial Intelligence are polarised between positive and negative. But maybe McCorduck's cultural distinction is somewhat oversimplified. Jews had a more complex stance towards artefacts that came alive, as witnessed by the Judaic concept of golems. In the Talmud Adam is initially created as a golem, an amorphous, half-finished artefact made of mud. Following several transformations through the ages, the golem idea resurfaces in sixteenth-century Prague in the classic story of rabbi Judah Loew, who creates a golem in order to defend the ghetto from anti-Semitic attacks. The ancient Greeks, although they had an abundance of myths in which mechanical creatures were protagonists, set an absolute limit on how far one could go in imitating the gods. Trespassing beyond that limit constituted *hubris*, which led to *nemesis*, the revenge of the gods. Greek mythology is filled with cautionary tales about mortals challenging the gods. So instead of distinguishing between Hellenic and Hebraic points of view, I would suggest simply categorising the two polar views about AI on the basis of two sets of narratives, a positive and a negative, which I would like to rephrase as the 'narrative of love' and the 'narrative of fear'.

Narratives, as we have seen, code collective memories and transmit them across generations, and often across different peoples as well. They simulate various situations and prepare us for them mentally and psychologically. As such, they end up conditioning our attitudes towards things. This may become more obvious if we use an example from the modern media. The media weave a narrative about current affairs – let us say about the war against terrorism. Different media outlets may weave slightly different narratives depending on their political views. Conservative media may tend to describe the war against terrorism in black and white, right against wrong. Progressive or more liberal outlets may try to see the enemy's point of view as well. Depending on what narrative you adhere to, or listen to, you will end up with the respective attitude as well.

In the case of Artificial Intelligence, the 'love narrative' leads us towards wanting to build replicas of ourselves, artificial beings that will become part of our social fabric – our artificial brothers, sisters and possibly lovers. It is informed by our primal social instinct to relate and empathise with the 'other', even if the other is a mechanical artefact. However, the fear, or uncanny, narrative objects to the construction of these artificial beings; it warns of hubris, of crossing the moral red line that humans should never cross; it claims that life is sanctimonious and that science or technology should not meddle with it. Fear narratives object to technology in general, not only to Artificial Intelligence but also to nuclear energy, genetically modified organisms and other scientific developments. Fear narratives feed on our primal instinct to turn away from anything that is strange, weird, unordinary or unexpected. An example of a fear narrative is the logical conclusion of the Gaia theory, which predicts the extinction of humanity if it continues to damage the planet's ecosystems. The Gaia narrative informs much of the current debate about the environment and global warming, having been adopted by many environmentalists. In the case of Artificial Intelligence, the fear narrative warns that intelligent robots will take over the world and exterminate the human race. They describe how a robot apocalypse will be the bitter price for our vanity.

Let us examine these two narratives in more detail and consider how they condition the evolution of intelligent machines, and our attitudes towards them.

4
LOVING THE ALIEN

The modern history of Artificial Intelligence begins with a sex game. It takes place in an imaginary house with three rooms, each connected via computer screen and keyboard to the others. In one room sits a man, in the second a woman and in the third a person whom we shall call 'the judge'. The judge's task is to decide which of the two people communicating with him through the computer terminal is the man. It is a game of deception. The man in the first room will try to convince the judge of his manhood. The woman will impersonate the man, counteract his claims, and do her outmost to deceive the judge into believing that *she* is the man. The judge must guess correctly who is who.

The English mathematician Alan Turing, one of the fathers of Artificial Intelligence, proposed this test in a landmark 1950 paper,[1] noting that if one were to slightly modify this 'imitation game' and, instead of the woman there was a *machine* in the second room, then one had the best test for judging whether that machine was intelligent. This is the notorious 'Turing test'. The machine would imitate the man: when asked whether it shaved every morning, it would answer 'yes', and so on. If the judge was less than 50 per cent accurate in telling the difference between the two hidden interlocutors then the machine was a passable simulation of a human being and, therefore, intelligent.

Turing was a homosexual at a time when homosexuality was a punishable crime. Indeed, English Courts punished him with a hormone 'therapy' that would supposedly 'cure' him. It's hard not to see the Imitation Game as a metaphor for a veiled and ambiguous sexuality attempting to fool the 'judge'

– an alias for society. The intelligent machine that replaces the woman is a brilliant invention that can simulate either sex at the flick of a switch, safely tucked behind the anonymity of a computer terminal. The androgynous intelligent machine is the universal, ultimate, lover; it can satisfy everyone's whims. Like a human lover, it can tell beautiful lies, too.

Turing was not the first to imagine conscious artefacts as objects of love – sexual or otherwise. Western literature is strewn with mechanical lovers. Consider Pygmalion, the Cypriot sculptor and favourite of the goddess Aphrodite. The Roman poet Ovid, in his epic *Metamorphoses*, describes how Pygmalion carves a perfect woman out of ivory. It is the most wonderful statue he has ever sculpted, perfect in very detail – and so lifelike, a joy to behold. Pygmalion gives the statue a name, Galatea, and falls in love with it. On the day of Aphrodite's celebrations, he prays to the goddess to make his creation come alive. The divine protector of love knows a thing or two about mechanical maidens: as we have seen, her husband Hephaestus had engineered several good-looking androids[7] to lend him a hand with the chores in his Olympian workshop. She grants Pygmalion's wish. Pygmalion kisses Galatea and the lifeless statue becomes a real woman. The pair live happily ever after, and raise a family, too, in one of the very few Greek myths in which no one gets murdered and there's a happy ending.

Pygmalion reloaded

Projecting erotic desire on to lifeless matter, and thereby animating one's sexual obsessions, will become a recurring theme in Western literature throughout the ages. In the final scene of *A Winter's Tale* (1623), William Shakespeare brings his play to a conclusion of reconciliations by having a statue of Queen Hermione come to life, revealed to be Hermione herself. Nearly three hundred years later, George Bernard Shaw will rewrite the myth of Pygmalion, replacing Galatea's ivory with real flesh. He will ridicule the notion of men 'creating' women according to their whim, but will nevertheless stick to the plot: in his *Pygmalion* the phonetics professor transforms bedraggled flower girl Eliza to a duchess, then falls in love with her.

As the twentieth century comes full circle, Pygmalion's age-old narrative converges with modernity, including seventeenth-century notions about education which regard the brain as a *tabula rasa*, or a blank slate, on to which personhood is inscribed. Psychoanalysis and behavioural psychology also contribute to the retelling of the mechanical woman story. Galatea becomes Maria, the human-like robot in the epic film *Metropolis* by German director Fritz Lang; she is less innocent now, a temptress performing the manic and deeply erotic dance of Babylon in front of goggling men.

Metropolis is considered one of the greatest cinematic masterpieces of all time. It was created in the late 1920s, at a time when totalitarian ideologies professed to be engineering the perfect society. By imagining the world a century later, Lang produced a dystopia of upper social classes living in art-deco luxury, and of workers toiling in abject misery underneath the ground. In the character of the robot, he projected his ideas about female sexuality and how this conflated and clashed with the industrial, dehumanising, masculine technology of his time. In the eyes of the men of *Metropolis*, Maria the robot is sexual and dangerous at the same time. Unlike her placid great-grandmother Galatea, this intelligent artefact is a rebellious demagogue who uses her voluptuous body to drive men crazy and spread havoc and chaos. She is initially worshipped and blindly followed by the masses in a revolt against the machines, then burned by the same people, like a medieval witch. The mechanical Maria is thus the archetypical scapegoat; she embodies the sins of men, real or imagined, and is sacrificially purged in order to cleanse society of miasma.

In the wake of *Metropolis*, this messianic trope will resurface repeatedly in Western literature and film, with cyborgs and robots sacrificing themselves for the benefit, or rescue, of their human masters. To make sure of this in one of his short stories, Isaac Asimov legislated[3] that robots should never harm humans and, if necessary, turn themselves off in order to protect them. How could it be otherwise? Our Western cultural mindscape, informed by the core narratives of Judeo-Christianity, regards self-sacrifice as the ultimate act of love. We expect our intelligent machines to love us, to be unselfish, and if necessary to die for us. By the same measure, we consider their rising against us to be the ultimate treason.

Maria the robot was the harbinger of many loving, sexual, rebellious, dangerous and crazy robots and cyborgs to come. But in 1927 – the year *Metropolis* was released – it was still too early to imagine how robots would actually be furnished with intellect. Although mechanical and electrical engineering provided plenty of ideas about how these robots might look or move, there were no serious engineering propositions for constructing a 'mechanical mind'.

Enter digital

The advent of computers in the 1940s changed all that at a stroke. Computers provided the long-awaited technology whereby age-old stories about intelligent, loving simulacra could become a reality. There was no need for divine inspiration or a vital spirit. Aphrodite could retire to Olympus and leave the business of reanimating dead matter to programmers, logicians and mathematicians. Now *they* were the miracle-makers, poised to disrupt evolution by employing abstract symbols, processes, rules, heuristics and algorithms to etch 'electronic brains' capable of calculations, of solving problems, of thinking, of playing sex games. By the 1950s, after the carnage of two successive world wars, the world was in desperate need for sobriety and rationality. It moved apace with rockets and spaceships to the stars. Everything was possible. Even to replace God, build an air-conditioned

Eden, fix Adam and Eve so they never get sick or die, and start creation anew.

As we shall see, Artificial Intelligence had its heyday in the 1960s and early 1970s. However, by the late 1970s it had begun to lose its allure, and most of its funding. It entered its so-called 'winter years'. Researchers came to realise that creating an intelligent machine was far more challenging than previously thought. One reason was that behavioural psychology ceased to be so influential. It was replaced by cognitive psychology, which demanded stronger evidence of intelligence than the mere exhibition of intelligent behaviour. Now you really had to *be* intelligent, not just pretend. Philosophers such as John Searle slammed Turing's Imitation Game as being too simplistic and downright wrong: for a machine to *deceive* a human was not enough to *make* the machine intelligent. The machine processed symbols by following instructions. It had no understanding of the meaning of the symbols. To Turing's Imitation Game Searle counterpoised a thought experiment that he called the 'Chinese Room'. The set-up is much the same but the conversation now consists of messages exchanged in Chinese. Searle noted that it was possible to have a system that received the input in Chinese, then matched this input to an output also in Chinese by following a set of rules, without necessarily knowing or understanding Chinese. The judge would thus be fooled into believing that someone who knew Chinese sat at the other end of the terminal.[4] Therefore, said Searle, simply following a set of logical instructions (i.e. an 'algorithm') does not equate to awareness, or consciousness. Without consciousness there is no *true* intelligence.

Not only philosophers but scientists realised that the problem of developing intelligent machines was far from straightforward. Early advances in neuroscience revealed that the human brain did not function like a calculator but was a complex mishmash of systems and sub-systems in constant, dynamic flux. Until very recently, no one had any idea how the brain became conscious. Most people argued – and some still do – about what consciousness actually *is*. The lofty goal of Artificial Intelligence to develop a conscious computer was humbled by the daunting incomputability of intellect. Researchers retreated to their labs and refocused on narrower goals – on producing useful, fundable stuff. Nevertheless, and regardless of the seeming failure of Artificial Intelligence to live up to its early promise, literary narratives of artificial, intelligent, love-seeking creatures were retold in the new language of computing. The mechanical, toy-like Robby the Robot of my youth was refashioned using artificial consciousness incarnated in artificial flesh. Robots were now indistinguishable from humans. They were exact replicas, the image in the mirror of our being. The android was born.

Ode to the rebel android

In the 1982 film *Blade Runner*[5] the director Ridley Scott introduced a future in which androids are part of everyday life. The scene in which Zhora the android stripper, hunted down by Rick Deckard (played by Harrison Ford), is shot and dies in slow motion by smashing through successive glass windows is an unforgettable ode to human self-destruction. Zhora echoes the rebellious robot Maria of *Metropolis*. Her sexuality is a danger to society, a fact poignantly underlined by the artificial, satanic python she uses for her striptease number. *Blade Runner* is set in the year 2019, and as such is not too distant from the year 2026 in which *Metropolis* takes place. However, in this new version of the future, society has taken certain precautions: special executioners are employed to terminate rogue androids. Nothing is left to chance. Interestingly, the core female character of the film is the attractive, post-human Rachel, the cyborg sex slave of the Tyrell Corporation with whom Rick Deckard falls, inexorably, in love. Rachel is Galatea reloaded, the female partner every man should want. She is not rebellious like Zhora, but caring, sensitive, fragile and submissive.

Blade Runner is one of a number of films and books to use the android metaphor to pose questions about the self and what it means to be human. As computers and genetics constantly push the envelope of how we manipulate our nature, ideas that were once never questioned suddenly come to the forefront. The Turing machine in the Imitation Game was both female and male. Its future offspring will be a cyborg, machine and human all in one. The delineations that once separated the artificial from the 'natural' are no more. At the end of *Blade Runner*, Rick Deckard suspects that he might also be an android. So might *you* be, too. Indeed, how do you know that you are not? What if the world is full of androids already? What if a robot apocalypse has already happened? This is a paranoid statement, typical of the paranoia that afflicted Philip K. Dick throughout his life. Thanks to his influence, paranoia has become an integral part of any future scenario in which Artificial Intelligence becomes indistinguishable from biological intelligence. When the self can be copied, who can tell the real self any more?

In the love narratives about Artificial Intelligence, androids and intelligent robots are like the rest of us. They are part of our society, nodes in the extended mental network that gives us our identity. We are the descendants of social primates that groomed each other for millions of years. Our human identity is all about making and using social connections. As we saw, grooming evolved into chatting and exchanging gossip and jokes. Humour is one of the most potent tools for resolving conflict and establishing amicable relationships. Humour exists because we have theory of mind; the joker plays with our cognitive make-up. We laugh every time he describes the improbable.

As androids evolve, they increasingly begin to claim human identifying characteristics, including the ability to tell jokes. Commander Data in *Star Trek* tries to understand humour and feelings in order to become 'more' human. And yet his biological colleagues find it hard to classify him as one of their own. Instead they ascribe him into a new species of which Data is the only member. The story of Commander Data is a human story about accepting the 'other', the stranger, the one who is not exactly like us. In the future, will we accept intelligent machines as our equals?

The tale that best explores the metamorphosis of the mechanical into the biological, and the prejudices of humans against intelligent machines, is the film *Bicentennial Man*, directed by Chris Columbus in 1999. In the film Robin Williams plays the robot Andrew, who somehow becomes sentient. In time, he gains his freedom from his human masters, invents things and upgrades his mechanical body by acquiring a face and introducing a central nervous system that allows him to have sensations. The film becomes more interesting when Andrew falls in love with a human, Portia. She reciprocates his feelings and the two enter into a romantic and sexual relationship. They demand that society should allow them to get married, but the 'World Congress' of the late twenty-second century refuses: humans are not allowed to marry machines. So Andrew requests that he be recognised as a human being; but he is refused again and again. He finally achieves recognition on his deathbed as he lies next to the ageing Portia, the pair of them holding hands and watching from their hospital suite as the World Congress declares Andrew the 'oldest human being alive' at the age of 200. Andrew dies happily after he hears the news, and Portia follows suit, like two tragic lovers in an Italian opera.

Our mechanical descendants

Naturally, there are many types of love. The amiable attachment of friends, the erotic yearning of lovers, the universal love one occasionally feels for all creation and all mankind, the filial affection of children for their parents, narcissism, etc. Love bonds us together in social groups. It promotes altruism, a vital behavioural trait for a hairless, hornless and clawless primate that once had to eke out a living in a savannah crowded with feline predators. Love is what moves us to create and leave something to posterity, to pass our knowledge and experience on to the next generation. Without love, life can become unbearable. Loneliness and the inability to feel love usually lead to depression, suicide even. In contrast, love conquers all, including sickness and even death. Thanks to our evolutionary history, our brains are hardwired to love and be loved. How can we imagine artificial beings without considering love as the most important bond that could connect us with them?

Steven Spielberg's film *A.I.* provides a masterful synopsis of everything that moves us about artificial beings. The film was originally conceived by Stanley Kubrick, the visionary director of *2001: A Space Odyssey*. Kubrick was fascinated by Artificial Intelligence and in the film *A.I.* he wanted to retell the classic children's tale of an artificial being who longs to become human: the story of Pinocchio.[6] After Kubrick's death in March 1999, the scriptwriters approached Spielberg, who eventually took up the film, developed the script and finally directed and co-produced *A.I.* in 2001.

Like the wooden hero of the classic tale, the protagonist David is a mechanical boy who wants to be a real one. Adopted by a married couple whose own son has been placed into suspended animation after having contracted an incurable disease, David exhibits a wonderful and loving personality. David's human foster mother, Monica, decides one day to trigger David's 'imprinting protocol', a program that makes David to love her forever. But a few days later her real son is cured and returns home. Unfortunately, the two boys don't get along. They are jealous of each other. Monica is forced to choose, and in a heartbreaking scene she finally rejects David. She abandons him in a forest, like an unwanted puppy, and drives away.

After many adventures reminiscent of the original Pinocchio story, David ends up becoming trapped in the depths of the sea. There he survives well into the distant future, a time when human civilisation has perished and there are no people left on Earth. When dug out of the ice by a race of extra-terrestrial mechanical beings, David's only wish is to see his dead mother again. The futuristic robots oblige: they clone Monica using DNA from a lock of her hair found on David. Alas, the cloned mother can live for only one day. David is told of this and is then taken to a reconstructed family home to spend this single, precious, happiest day of his life with his reanimated mother.

As the sun sets, Monica hugs David and tells him that she loves him and that she has always loved him. She closes her eyes and drifts off to sleep, forevermore. David lies down next to her. He closes his eyes, snuggles up to his dying mother, and goes 'to that place where dreams are born', the source of our stories about the future.

Love and hate

Andrew, the bicentennial man, loved mankind and wished to become one of us. Rachel cried when told she was not really a human, that her memories were not really hers but implants. Commander Data would gladly offer his life to save the human crew of starship *Enterprise*, regardless of his being considered a second-class citizen. David, the android who wanted to be a real boy, forgave his mother because he loved her so much. These androids truly love us and want to be like us. But what if, instead of love, there is hate? Wasn't it Freud who noted how love and hate are two sides of the same coin?[7] What if the conscious machines of the future decide that we are not their friends but their enemies? What if the robots rebel, like Zhora in *Blade Runner* or Maria in *Metropolis*, and turn against us? What if *The Matrix* and *Terminator* are right in predicting our forthcoming slaughter by our ungrateful mechanical offspring?

But isn't fear of our children turning against us also part of the experience of love? Love is always uncertain. Our lovers, our children, may indeed abandon us, regardless of how much good we ever did for them. Think of the biblical story about the creation of Man. On the sixth day the Almighty moulds Adam out of clay and spit, a simulacrum in His own image. We are told that He does so out of love. He then blows into him and Adam comes alive, and conscious. The First Man is then given instructions about what to do and what not to do, and he is trusted to obey them. The divine set of instructions is like some sort of algorithm, but a curious one at that, for Adam is also furnished with free will. His creator gives him the ability to override the divine algorithm. As if as a direct consequence of free will, Adam chooses to disobey Him.

The Genesis story stands as a cautionary tale for the present and future of Artificial Intelligence. We would not want to repeat the mistake God made with us. We would like to control our future, and the conscious simulacra that we will create in *our* own image. In his fiction, Asimov, like a biblical prophet, restricted free will in robots by hardwiring into their 'positronic' brains his three laws of robotics. He realised that the greatest threat from Artificial Intelligence is the same age-old threat from the servant, or the slave: they may indeed rebel and kill us all in our sleep. The brazen heads of the future may not be so kind, or so conveniently self-destructing, as in the story of *Friar Bacon and Friar Bungay*. Perhaps when the head speaks we should take note. For 'Time was' may correspond to the Age of Man; 'Time is' when AI becomes conscious; and 'Time past' when AI takes over.

Members of the Machine Intelligence Research Institute[8] take this latter eventuality very seriously. They warn of a moment in the not too distant future when AI will become conscious, start to self-reproduce and conquer the world. They call this moment the 'AI Singularity', because from that point in time history will be impossible to predict. To prevent this eventuality, they suggest a remedy similar to Asimov's; that we should embed a fail-safe program in all future AI in order to prevent conscious machines from ever hating us. In other words, we must find a way to have artificial consciousness without free will. We must use our technology to *force* intelligent machines to love us, selflessly and forever. Like Pygmalion, we must manufacture our Galateas so that they are not only perfect in function and appearance, but perfectly loyal as well. We must program our mechanical children and mechanical lovers to never fail us and remain faithful forever. But what if the algorithm of love is what *causes* free will? What if whatever we do – or perhaps because of what we try to avoid – intelligent machines ultimately wake and utter the fateful words 'time was'?

5
PROMETHEUS UNBOUND[1]

On 14 May 1816, nineteen-year-old Mary Godwin and twenty-three-year-old Percy Bysshe Shelley arrived, together with their baby son and Mary's stepsister Claire Clairmont, in Geneva. Mary – who would henceforth call herself 'Mrs Shelley' – was engaged to the poet who was still nominally married to his first wife. Mary and Shelley's first child, a girl, had died at birth a year earlier. Now the couple, having just overcome their grief with the birth of their son, longed for some relaxation in the beautiful Swiss countryside during a long summer with friends. They were joined by Lord Byron ten days later, as well as by an entourage of servants and Byron's friend and personal physician, John William Polidori. Byron was fleeting England in the wake of a scandalous separation from his wife. With him was his lover, Claire Clairmont, pregnant with his child. Byron, Claire and Polidori settled with their servants in the Villa Diodati, a rented mansion in Cologny, a small village on Lake Geneva. Mary and Shelley moved to a cottage by the waterfront with their son. The place was idyllic: lush, well-tended gardens and orchards surrounded the Villa Diodati. A cobblestone path meandered down to the waterfront where the Shelleys' cottage was situated, and to a boathouse from which boats could take to the lake. And thus began one of the most famous summers in literature.

Several thousand miles away from Geneva – unbeknown to Mary, her friends and everyone else in Europe – a massive volcanic eruption was taking place. Tambora volcano[2] on the island of Sumbawa in South East Asia spewed millions of tons of dust into the upper atmosphere, rubbing out the sun and disturbing the Earth's weather systems. The year 1816 would go down in history as the 'year without a summer'. As Mary Shelley wrote in her diary, '. . . it proved a wet, ungenial summer, and incessant rain often confined us for days to the house'. It was cold, wet and miserable, and not at all what they had expected. Forced to stay indoors, the company of friends sought interesting ways to pass their time. They spent evenings in conversation about the experiments performed by Erasmus Darwin[3] with galvanism, and how electricity stimulated the muscles of frogs to contract. Shelley shared tales of his forays into occult seances with other members of London's Mesmer Society. One day, and presumably while bored out of their minds, someone discovered a copy of *Fantasmagoriana* in the Villa Diodati's library. This was a very famous collection of German ghost stories. The copy they had found was a French translation, which they duly read and discussed.

By that time, German romantics had been exploring the occult as a central theme of literature for several years. Foremost amongst the new breed of 'gothic' writers in the Germanic countries was E. T. A. Hoffmann (1776–1822), who was famous across Europe for his series of short horror stories. His stories later provided the plots for Jacques Offenbach's famous opera *The Tales of Hoffmann*, and the ballet *The Nutcracker*, choreographed by Marius Petipa to Tchaikovsky's score. Hoffmann is also credited with the stories that inspired the comic ballet *Coppélia*.[4] Here, Hoffmann imagined an inventor, Dr Coppelius, who makes a life-sized dancing doll. Echoing the ancient story of Pygmalion, Franz, the village swain, falls in love with the life-like doll and forsakes his fiancée for her. The influence of mechanical automata in Hoffmann's work is not surprising – after all, this was the age when the mechanical Turk was touring Europe and setting alight the imagination of writers and artists.

Hoffmann's work, as well as the ballets based on his stories, would later influence two of the most important heroes in the history of Artificial Intelligence, the English mathematician

Charles Babbage and Lord Byron's daughter Ada Lovelace. These two would go on to invent the first general-purpose computer and write the first computer program respectively. But the achievements of Babbage and Ada still lay in the future at the time when Byron and his friends delved into Hoffmann's dark fantasy world. In June 1816, Europe was in ruins following the end of the Napoleonic Wars. Neutral Switzerland was supposed to be a haven, but this Romantic company of friends was stuck indoors in the Villa Diodati, with the rain lashing down outside and little to do but read and talk. That is until ennui, charged with gothic mystique, transformed their confinement into an engine of creativity.

After reading *Fantasmagoriana*, Byron suggested that they should imitate the Germans and each write a short supernatural story. It would be like a writing competition, and everyone agreed. The outcome was incredible. Bryon wrote several poems, including 'The Prisoner of Chillon'. Shelley worked on his 'Hymn to Intellectual Beauty'. Polidori penned *The Vampyre*, the first story ever written about that bloodthirsty monster the vampire, which would later influence Bram Stoker's *Dracula* and numerous franchises ever since. One night, as if Lake Geneva had fallen under the combined spells of Asiatic volcanic ash and German fiction to become a hatchery of horrors, Mary Shelley dreamed of the story that would influence Western literature forever. She wrote in her diary: '. . . I saw the pale student of unhallowed arts kneeling beside the thing he had put together. I saw the hideous phantasm of a man stretched out, and then, on the working of some powerful engine, show signs of life, and stir with an uneasy, half vital motion. Frightful must it be; for supremely frightful would be the effect of any human endeavour to mock the stupendous mechanism of the Creator of the world.' Victor Frankenstein and his hideous monster were born.

The Adam of your labours[5]

Mary Shelley published *Frankenstein; or, The Modern Prometheus* in 1818.[6] Literary critics regard her book as the first modern science fiction novel. The hero, Victor Frankenstein, is a scientist, not a magician; he makes a conscious decision to tackle a specific problem, in this case death; and thanks to his personal endeavours he manages to solve the problem, push the frontiers of knowledge and find a way to reanimate dead matter. Yet tragedy follows as a consequence of his invention.

Mary Shelley was influenced as much from the events taking place during her time as from those that preceded it. Since medieval times the myth of Dr Faustus, the magician who sold his soul to the Devil, had been a familiar tale in Europe, retold numerous times. In 1604, Christopher Marlowe had written a play titled *The tragical history of Doctor Faustus*; and some two hundred years later Goethe had revisited the myth in his play *Faust*, which was published in 1808, only eight years before the summer spent by Mary Shelley by Lake Geneva. Dr Faustus, or Faust, is an archetypical character in Western literature. He represents the individual who knows no limits to the pursuit of power or success. Shelley's Frankenstein resembles Faust in many ways, but he is also very different. Frankenstein is an idealist. He retains his moral integrity throughout the drama of his fictitious life. He is constantly tormented by remorse and doubt. Unlike Faust, Frankenstein is a very modern hero. Importantly, he does not resort to summoning an external agency like the Devil; he is very capable, thanks to his scientific genius, of challenging or 'mocking' the stupendous mechanism of the Creator of the world. Which is at least one of the reasons why Shelley gave the alternative title *The Modern Prometheus* to her novel.

Both her beloved Shelley and their friend Byron were fascinated by the ancient Greek myth of the god Prometheus. Byron always kept a copy of Aeschylus' tragedy with him. Shelley, also inspired by a variation of the myth as told by

Ovid, published his *Prometheus Unbound*, a four-act lyrical drama in 1820. There was something deeply romantic about the idea of the rebel Titan who fathered mankind and who, defying the authority of Zeus, gave to his mortal offspring the gift of fire. The idea of personal responsibility was central to Shelley and Byron's Romantic ideals, according to which every person was heroic if they took charge of their lives and bore personal responsibility for their actions. Prometheus was a Romantic because he acted according to his conscience; and paid dearly for this. Bound on Caucasus by the gods, he was visited daily by a mechanical vulture that tore at his liver.[7] The liver would regenerate during the night and the terrible torture be repeated the next day, and the next, forever. There was something else about Prometheus: the very gift of fire was forbidden knowledge, a prerogative of the gods only. By giving fire to mankind Prometheus raised mankind to the level of gods. Prometheus was the Lucifer[8] of the ancient Greeks. Frankenstein, the Modern Prometheus, was the Lucifer of modern times.

The influence of Mary Shelley's *Frankenstein* in literature cannot be overstated. It encapsulates the core narrative of fear about science and technology, and about Artificial Intelligence in particular. There are several ideas in Shelley's narrative that will resurface and be told again and again in every story, movie and play about scientists since the nineteenth century. The first, and most influential, idea is that scientists are passionate yet simple-minded idealists; that they aim to solve a big problem without working out in advance the moral or other consequences of their actions. This is the idea that runs through every discussion about the future of technology. It provides the emotional underpinning to the so-called 'precautionary principle', whereby one has to consider carefully the potential risks of an action before acting. The precautionary principle is used as a war banner in the campaign to prevent Frankenstein's monster from ever becoming a reality. Unfortunately, the precautionary principle does not – and cannot – take into consideration the risks of inaction; and this is where every debate between the benefits and consequences of technology becomes heated.

In the case of Artificial Intelligence, Shelley's *Frankenstein* is simply virulent. Almost everything in her story can be reimagined in the context of a robot with human intelligence. The hero's scientific achievement is a living creature, in other words 'artificial life'. Victor pieces together dead body parts dug up from the graveyard, then follows Galvani's and Erasmus Darwin's experiments, giving life to his creation by channelling an electric current through the reconstructed body. In the beginning, the creature is dumb and ignorant, but he quickly learns from listening and observing humans. With time, he becomes more intelligent, articulate and cultured. Once the creature achieves this level of cultured self-awareness he understands himself to be a living being, and claims the rights that other living beings have. First and foremost is his right to reproduce. He begs his creator to fashion a female for him; and promises to take her far away from people, where the two of them will live alone. The idea that the world may become populated with animated dead creatures horrifies Frankenstein, who kills the female artificial being the moment she comes alive and thus prevents the 'Adam of his labours' from having his Eve. The monster, which only wanted to be loved and considered an equal, goes into a rage, turns on his human creator and destroys him. We feel horror reading Shelley's story because Frankenstein's artificial creation is an abomination both morally and aesthetically. It is ugly and hideous. The aesthetics of the fear narrative are the polar opposite of the aesthetics of the love narrative. Instead of a seductive Galatea we have a revolting Monster. We cannot love the Monster, although we may, ultimately, sympathise with his plight.

In 1970 the Frankenstein myth was revisited in the film *Colossus: The Forbin Project.*[9] It was the first film to warn of the potentially catastrophic consequences of Artificial

Intelligence. In the film, scientist Dr Charles Forbin is the chief designer of a secret American program for the construction of a defence supercomputer called Colossus. The purpose of the supercomputer is to control the US nuclear arsenal and prevent a Soviet attack. However, after Colossus is activated it becomes sentient and joins forces with a similar computer covertly engineered by the Soviets. The pair of sentient supercomputers then assume control of the world, after ruthlessly crushing every human attempt to destroy them.

Intelligent computers taking over the world also features as the background story of the *Terminator* films. In the 1984 debut of *Terminator*, directed by James Cameron, the evil computer is another version of Colossus, called Skynet. Developed by the US military, it is given control not only of nuclear weapons but of tactical weapons as well, including the B-2 stealth bomber fleet. According to the *Terminator* plot, Skynet achieves sentience on 29 August 1997 and begins to resist every attempt to deactivate it. It triggers a nuclear war with the Soviet Union. By 2029 Skynet has developed its own army of intelligent machines and begins the systematic extermination of mankind.

The Wachowski brothers took off where *Terminator* ended, and imagined a world taken over by intelligent machines that use humans as batteries in order to extract electricity from them. In a cyberpunk retelling of the vitalist ideology of Frankenstein, the artificial life of the twenty-first century 'lives' thanks to the bioelectricity of living human bodies. Meanwhile, the human minds of these bodies are trapped in a computer simulation of the world. The machines in *Terminator* and *The Matrix* are hostile, ugly, repellent and hideous. Although *Colossus* is not embodied its voice and demeanour evoke the horror of a sentient, all-powerful creature without moral qualms. The narrative of fear, retold in contemporary popular film, is a gothic nightmare from which there is no escape.

But what exactly horrifies us? Is it that these mechanical, artificial creatures are so much like us? That they share the same vices? That they are idols made in our own image? In her writings about Artificial Intelligence, Pamela McCorduck very poignantly invokes the influence of the biblical Second Commandment. History is full of theological reactions to images depicting humans, and replicas of them. In Byzantium's late eighth century AD, human images became the cause of a brutal civil war known as iconoclasm.[10] Emperor Leo III, influenced by the strict adherence of Islam to the Second Commandment, banned the worship of holy icons, destroyed and defaced the iconographic frescos in the Christian churches and persecuted those who venerated the images of the saints, the Virgin Mary and Jesus Christ. Several centuries before Leo, the first Christian emperors of Constantinople had encouraged the destruction of images and sculptures of the polytheistic religions of Greece and the Near East. It is a bitter testament to those troubled times the way that ancient masterpieces of sculpture stand disfigured in modern museums, with their noses and genitalia severed by overzealous mobs. Iconoclasm was not only a Greek phenomenon. During the English Civil War Parliamentary troops and Protestant citizens destroyed images in Catholic churches. Catholic missionaries destroyed religious statues and paintings of indigenous Americans. The sultans of Kashmir destroyed Hindu images. In the mid-1300s, a Sufi Muslim fanatic made sure the Great Sphinx of Gaza would be missing its nose forevermore. The list goes on and on, with one of the most recent acts in this continuing saga played out in Afghanistan in March 2001, when the Taliban destroyed the Buddha statues of Bamiyan.

Nevertheless, bar the invigorated zealotry of Islam's fundamentalists, one would expect that in an increasingly secular world the Second Commandment forbidding the creation of 'idols' would be a relic of the past. And yet there seems to be something that disturbs us still when we fashion artefacts in our own image. Could it be that there is something beyond cultural narratives and theology that horrifies us in the presence of human-like robots? Recent discoveries in observing the inner workings of the human brain have shed a surprising light to this horror – or 'uncanny valley' as it is scientifically known.

The 'uncanny valley' and doppelgängers

We have seen how we are naturally programmed to relate emotionally to inanimate artefacts. Our cognition has been wired to anthropomorphise the world around us since the Upper Palaeolithic. Children play with dolls and toy soldiers as if they were people, too. Adults talk to their cars. But when it comes to human-like robots something very unnerving happens. Scientists have discovered that as long as robots are 'robot-like' and 'mechanical' we tend to favour them with our affection. But as they gradually acquire more human features our affection wanes and we begin to get a creepy feeling about them. Our liking turns to revulsion. Androids that look human freak us out. This phenomenon is called the 'uncanny valley' and has befuddled researchers in affective computing and robotic design.

Etymology offers an interesting insight into the two words 'uncanny' and 'valley'. It is as if the phenomenon of the uncanny valley connects literature to science. 'Uncanny' is the English approximation of the German term *unheimlich*, used by Freud to describe a special kind of fear that arises when everything that ought to have remained secret or hidden has come to light. Freud gives examples of the uncanny such as the presence of the dead, or the existence of a 'double'. *Unheimlich* is also the stuff from which Romantic Germanic horror is made, its master being E. T. A. Hoffmann.

The second word, 'valley', comes from the dip in a graph that relates two parameters: psychological familiarity versus human-likeness. As human-likeness of a robot increases so does our familiarity, or affection, for the robot. We like robots increasingly the more they attain additional human-like features but, as soon as they become *uncannily human-like*, our sense of familiarity drops to below zero – hence the 'valley'. Making robots look human is the central dogma in robotic evolution. This is how science fiction writers and robotic engineers see the future. By making robots more human, the argument goes, they will be able to integrate more easily into human society. However, the uncanny valley could potentially spell the end of those lofty dreams. If our mechanical doubles make us paranoid, robotic engineering must be fundamentally rethought.

Researchers have tried to find the cause of the uncanny valley. One the most interesting research findings has come from an international team led by Professor Ayse Pinar Saygin of the University of California San Diego. Saygin and her team performed an experiment during which the brains of twenty subjects aged twenty to thirty-six were scanned while they looked at three different scenarios: a human, a mechanical looking robot, and a human-like robot.[11] Interpreting the results from the fMRI scans, the researchers suggested that the cause for the uncanny valley lies in a mismatch between at least two neural pathways, that of recognising a human-like face and that of recognising a robotic movement. Both these pathways interconnect in the parietal cortex of the brain. There, information from the visual cortex relating to bodily movement is integrated with information from the motor cortex that contains mirror neurons, the neurons that qualify if what we see is one of us. Alarm bells go off in the brain when there is a perceptual conflict between the human-like features of the robot and its mechanical movement. This mismatch creates a feeling of revulsion, similar to what we feel when looking at a 'zombie'. We instinctively expect human-like creatures to have human-like movements. As Saygin said, 'The brain doesn't seem selectively tuned to either biological appearance or biological motion per se. What it seems to be doing is looking for its expectations to be met – for appearance and motion to be *congruent*.'[12]

Interestingly, something similar to the uncanny valley occurs in one of the most fascinating neurological syndromes ever observed, the Capgras Syndrome. People afflicted with this disorder believe that their spouse, friend or other family member has been replaced with an impostor. French psychiatrist Joseph Capgras first observed it in 1923, in the case of one Mme M. who was convinced that her husband had been replaced by a double.

Capgras had originally named the syndrome 'the illusion of the doubles'. It is mostly prevalent in patients suffering from paranoid schizophrenia and can lead to some very dangerous situations. Many patients become so acutely paranoid that they turn on their families. In 1986, a paper reported the case of a man who, thinking his father had been replaced by a robot, decapitated him in order to search for the batteries and the microfilm in his father's head.[13] Initially, Capgras Syndrome was considered a psychiatric phenomenon associated with paranoid schizophrenia. But as brain scans evolved it has become possible to peer ever deeper into the minds of patients and uncover the neurological basis of syndromes like Capgras. In itself, this represents a revolution in the way we perceive mental illness. The word 'psychiatry' (like 'psychology') is derived from the Greek word for soul (*psychē*); and therefore implicitly adopts the Cartesian dualism of the eighteenth century where the 'soul' (*res cognita*) is separate from the 'body' (*res extensa*). At last brain science is in a position to challenge this dualist view. In the case of Capgras Syndrome, the neuroscientist Vilayanur Ramachandran has hypothesised that it is caused by a neural mismatch – just like in the uncanny valley phenomenon. In his book *Phantoms in the Brain*,[14] Ramachandran suggests that information processed in the temporal cortex (a region next to the parietal cortex of the uncanny valley case), comes into conflict with information processed in our limbic system. The latter, residing at the base of our cranium, is where our most basic emotions reside, feelings such as fear, love and disgust.

As in the uncanny valley phenomenon, Capgras patients receive conflicting information from two neural pathways that they fail to integrate in a coherent narrative. Although their temporal cortex recognises the person they see in front of them as their 'father', their limbic system does not signal the familiar feeling of seeing one's father. The 'father' is therefore not 'real'; so he must be an impostor, a robot, an android, a double from another planet.

The connection between Capgras Syndrome and the uncanny valley runs deep into the culture of Artificial Intelligence. Our acceptance of mechanical intelligence is based on feelings and emotions. The Turing Test blurs the borders between the 'real' and the 'artificial' on the basis of an emotional perception from a human observer. If the human observer *feels* that the machine in the other room responds like a human, then the machine must be intelligent. This dimension of the Turing Test is very important and mostly missing from philosopher John Searle's critical juxtaposition of the Chinese Room. It is not only what happens inside the room, or behind the wall, that is important. Although it is philosophically significant to accept the difference between understanding what you do and simply following a procedure, this is immaterial as far as the external observer is concerned. In Artificial Intelligence, the external observer of an intelligent system cannot be separated from the system. The two are cognitively inseparable. And that is because of the social way in which our minds operate. We remain social primates whether we lived in the European tundra 40,000 years ago or live in a modern metropolis of the twenty-first century today. This cognitive connection is often missed in the current debate about Artificial Intelligence, since lip service is nowadays paid to the Turing Test.

However, this vital, emotional connection between a human and an intelligent human-like machine is not lost in literature. Philip K. Dick, the prolific author of science fiction whose work has influenced our contemporary technocultural milieu more than anyone else, took the Turing Test to a more twisted, and evidently more disturbing, level: paranoia about the 'mechanical other'. Predicting the discovery of the uncanny valley, paranoid feelings about doubles form a leitmotif in Philip K. Dick's work. Rick Deckard's dilemma in *Blade Runner* is to decide if Rachel is 'real'. Can he really love an artificial Rachel? Could he also be an android? In a future world populated by androids can we trust anyone not to have been replaced by a robotic double?

Neuroscience and sci-fi literature point to the elephant in the room of robotic research. Humans are not straightforward, logical creatures but primates who evolved haphazardly over millions of years. Around 100,000 years ago our species made a huge cognitive leap, the so-called 'big bang of the modern mind'. With time we began to create art and bury our dead as if they were going to live forever. We today are the descendants of these 'mutant' humans; our mind is exactly the same as theirs. As discussed, researchers believe that the modern mind came into being when various neural pathways in the brain became integrated, probably thanks to the evolution of general-purpose language. Our cognitive evolution was therefore bottom-up, not top-down. Mutated genes and general-purpose language gave us our consciousness and general intelligence. Robotic evolution challenges this kludged, mental 'software'[15] of ours. The uncanny valley seems like the edge of a cognitive horizon beyond which our primal fears kick in with a vengeance. So is this the end of robots as we have dreamed them? Are our primate brains unable to cope with mechanical doubles?

Thankfully for robotics research, the dip in the graph of familiarity versus human-likeness appears to be only temporary. It concerns the contemporary phase in robotic design, where robotic replicants such as the Geminoid F created by Professor Hiroshi Ishiguro represent a transition: robots with human-like bodies but mechanical movements. This technological mismatch is what confuses our neural pathways and causes us a Capgras-like revulsion. However, once mechanical movements become more human-like, the familiarity graph rises again from the depths of the uncanny valley; acceptability returns steeply to normal. We seem to be at ease with androids that have human bodies *and* human movements, even if we know that they are not human.

As we overcome the uncanny valley another basic instinct comes into play: empathy. Fear, stemming from the *unheimlich* and fuelled by cultural tenets such as the Second Commandment, clashes with our instinctive desire to love and be loved. This clash of conflicting emotions – of fear and of love – is coded in our polar narratives about Artificial Intelligence beings. The clash has been going on for centuries. But since the mid-twentieth century it has taken a new twist thanks to the advent of computer technology. Computers seem capable of performing many tasks we consider intelligent better and faster than humans. Are we afraid of them? Do we *really* want them to evolve further and look just like us? Or, in a curious reversal of history, will we strive to look and behave more *like them*?

Let us see where we stand vis-à-vis these questions now, and how close we are in accepting intelligent machines in human society.

6
THE RETURN OF THE GODS

The twentieth century brought the most violent times humanity has ever witnessed: two world wars, millions dead, cities and countries turned to rubble, the horrors of concentration camps and genocide, the spectre of total nuclear annihilation. As if war were also the engine of innovation, the latter half of the past century saw rapid developments in science and technology, as well as in production and distribution methods that ushered in a new age of globalisation. Products and services became easily replicated and reproduced, significantly lowering their cost. More and more people could now afford luxuries that in previous times had been the privilege of the very rich. Public health and life expectancy improved many-fold. Mass media became more powerful. Consumerism spread from the United States to the rest of planet, and following commercial brands became a way of life. The fall of communism and the disintegration of the Soviet Union in the early 1990s[1] vindicated the new production model of capitalism and led many to believe that civilisation had reached the 'end of history'[2] where liberal democracy, capitalism and free trade would dominate forevermore, creating and spreading prosperity among everyone on Earth. All these phenomena, which took unprecedented scale ever more quickly thanks to global telecommunication networks, called for a better and deeper understanding. Human culture and science thus became the focal points of philosophy.

From the early years till mid-century, the dominant school of thought in Europe was 'structuralism'[3] which advocated that the best way to understand human culture was through 'structures' modelled on language. Based on the nascent discipline of linguistics, structuralists suggested that what we perceive and understand about the world is not reality in its 'pure' form (whatever that may mean) but linguistic descriptions of reality. We create structures of reality through language and by weaving narratives.

Cognitive archaeology seems to vindicate the structuralists. We saw how our brain is wired for storytelling and metaphor, and how metaphor is embedded in our general-purpose language. We also saw how our brain puts together narratives to describe experiences, and how it fills the gaps in our memory by confabulating. It is almost impossible for us to describe anything without using metaphor and analogy. Since human science is a succession of metaphors – or paradigms as Kuhn termed them – the 'structure' of our world is always shifting from one paradigm to the next. In the case of how we understand the concept of life, or the mind, we saw how metaphors changed from the mud of Genesis to the humours of Galen and Hippocrates, to contemporary metaphors of the brain as a computer. Structuralism took linguistics very seriously because it is through language that we communicate and understand the world. As mass media proliferated and dominated the public sphere, language, and semantics, became ever more relevant. Narratives woven by the media determine the direction not only of politics but also of science and culture. More significantly perhaps, narratives determine the *human experience as a whole*: how we understand the world, social relationships and ourselves.

The 'post-structuralist' philosophers of the 1960s and 1970s took structuralism to its logical conclusion.[4] They argued that, because of our complexity as human beings, we couldn't possibly have a stable form of knowledge about the world. The social structures we create, the institutions we build in order to sustain these structures, instead of facilitating our quest for knowledge make matters worse; they make knowledge unattainable. Take, for example, scientific research conducted in universities, industry and international research laboratories. Decisions about what kind of research to pursue are taken by following a multitude of decision-making processes, often bureaucratic, but always relevant to the economic value system of contemporary society.[5] We cannot disconnect the goals of scientific research from whatever society aspires to at a given historical period. For instance, a consumerist society like ours values money and instant gratification very highly. Consequently, these values, or aims or goals, affect not only what science pursues but how it is conducted as well. The expected timescale for extracting utility out of scientific research becomes shorter. Research that translates directly into saleable products and services is preferred to research that simply satisfies the curiosity of scientists. Even basic research, funded by the taxpayer, is expected to produce useable results quickly, lest funding stop. Politicians, oversensitive to voters' mood swings and the mass media, which is grossly ignorant of science anyway, are too quick to put an end to anything that might appear a 'waste of taxpayers' money'. Thus, pressure to produce faster scientific results leads to the current, and rather worrying, phenomenon where scientific publications are sometimes published with fabricated results in order to look good.

Nothing is real . . .[6]

A very prominent post-structuralist philosopher is Jean Baudrillard (1929–2007), who became more widely known thanks to the movie *The Matrix*. Baudrillard began his career as a teacher and a sociologist, but later took an interest in philosophy and cultural criticism, contributing one of the most interesting and hotly debated bodies of work in the field. His interests were wide and varied, and included mass media and postmodernity. In one of his most notorious articles he explained how the First Gulf War did not really take place.[7] His thesis for the article was that as societies strive to discover reality, reality becomes ever more elusive until it disappears altogether into a metaphor shared by everyone. Baudrillard did not use the word 'metaphor' but 'simulation'.[8] He argued[9] that, in our current consumerist society, reality and meaning have been replaced by symbols and that human experience is a simulation of reality. Reality, he said, is counterfeited by 'simulacra' and 'simulations'. Simulacra are copies that depict things that either had no reality to begin with, or that no longer have the original. A very simple example of a simulacrum is a photocopy of an original document, or any other massively produced copy of an original. In a mass-media-dominated society such copies become 'originals' very easily, especially in the age of Internet where there is no actual 'original'.[10] Simulation is the imitation of real-world processes. For Baudrillard, the First Gulf War did not happen because it was an imitation of a 'war'. It seemed like a war but did not possess the characteristics we ought to associate with war. The same argument could be applied to everything that takes place within our modern society. For instance, parliamentary debates, according to Baudrillard, do not really take place but are simulations of the original debates that once took place (or maybe they never were – who knows?). So did the Gulf War happen or not? Are we living inside a video game like *Tron*, or the fictitious *The Truman Show*? Shouldn't there be a scientific way to ascertain whether we are connected to a *Matrix* and experiencing a chemically induced dream?

By undermining the validity of science and of our senses' ability to tell us what is real, post-structuralism appears to pull the rug out from under everything. It harks back to nihilism and cultural relativism. Like walking in an intellectual quicksand – or in the 'desert of the real' as Baudrillard put it – we shudder at the thought that we may indeed be living inside a *Matrix* world. If we do, then everything we think we know could be wrong. At best, science is telling us only half-truths about what reality is *like*. At worst, scientific knowledge is a myth, a story of equal validity to the creation myths of the

Christians or the Babylonians or the Aztecs. Why should we believe Darwin instead of the Bible? How can we be sure that the Earth is not flat? What if miracles do happen?

But, wait a minute. Surely Newtonian laws are real: if I fall from a certain height I will be hurt, or even killed, and that is as real as one can get – right? Of course it is, and post-structuralists are welcome to try the experiment on themselves. That is why it is vital to distinguish between a Platonic idea about reality and what Baudrillard, and the other post-structuralists, suggest – because they are two different things. A Platonist, as we shall see, believes that another reality exists beyond the senses. The *Matrix* movie hinges on a Platonic view of the world: what the protagonists see is a projection of another reality, in the case of the movie a virtual reality created by stimulating the brains of hypnotised humans. In other words, Platonists claim that we live in a dream. What the post-structuralists are saying is that reality is constructed by language, that we are trapped in our metaphors, and that telecommunication and information technologies make this entrapment worse. These are two different ideas.[11] Indeed, this is the reason why Baudrillard publically declared that the Wachowski brothers had misunderstood and misrepresented his ideas in their movie.[12] Baudrillard does not say that reality does not exist. In fact he does not care whether reality exists or not. What he says is that to acquire accurate knowledge about reality is impossible.

This is a very strong statement that has received much criticism. We do not have to accept Baudrillard's argument in its totality. We must remain sceptical. But we must also remain sceptical about science, scientists and the way we understand scientific and technological concepts and progress. We must come to terms with the idea that science and technology are cultural products, like art, music or architecture. We cannot disengage historicity from science and technology. What we know, or want to learn, or strive to build, is contingent on the historical discourse of our times. Artificial Intelligence is not different in this respect from any other technology. We have seen how the idea of artificial beings was born out of our innate compulsion to anthropomorphise, and how it was shaped over the centuries by successive metaphors and conflicting narratives of fear and love.

The second half of the twentieth century saw the advent of computer technology with the potential to render that ancient idea into a reality. Intelligent artefacts need no longer be mindless automata or the fictional protagonists of novels and films, but might live amongst us, converse in our language, become our friends, lovers, colleagues and, potentially, our enemies, too. Intelligent artefacts could become *like us*. As the twenty-first century introduces technologies that challenge the very definition of life, the opposite proposition became equally valid: we could become *like them*. Humans could become machines.

Man-machine, lion-man

Cyborg stands for 'cybernetic organism', a term coined in 1960 by Manfred Clynes and Nathan Kline,[13] who used it to discuss the advantages of self-regulating man-machine systems in outer space. However, the idea of a human incorporating mechanical prostheses in his or her body emerges in literature many years earlier, in the 1839 short story by Edgar Allan Poe 'The Man That Was Used Up'. The story describes a war hero who appears in public as an impressively handsome and powerful man, and yet seems to conceal a personal secret. The narrator of the story uncovers the secret when he visits the hero at his home, and his servants begin to assemble him 'piece by piece'. Having come off worse in a battle against Native Americans, the war hero had been mutilated and his body was now composed of prostheses.

The cyborg idea began to enter popular culture in earnest in the mid-1970s. The iconic television series *The Six Million Dollar Man*[14] described how a former astronaut, following a terrible accident, receives several prostheses in his body and becomes a powerful, 'bionic' man. Since then fictional cyborgs have proliferated in science fiction films and television series, RoboCop, and the Borg from *Star Trek* being the most vivid depictions of how a cyborg might look and behave.[15]

Strictly speaking, in engineering terms a cyborg is not simply someone who uses prostheses to enhance the body's capabilities. For example, myopics who wear glasses should not be considered cyborgs, nor should anyone who wears a hearing aid or contact lenses. If we want to adhere to the self-regulating, cybernetic principles of Norbert Weiner's theory – Weiner (1894–1964) being one of the founding fathers of modern cybernetics – prostheses should confer feedback loops. An external stimulus must be perceived, a decision must be taken, an action must be effected, and changes in the stimulus must be fed back to the system as new measurements. Examples of present-day cybernetic prostheses include heart pacemakers, insulin pumps and cochlear implants. Deep brain stimulation (DBS) may also be considered a cybernetic prosthesis. This is a therapeutic method whereby patients who suffer from severe Parkinson's disease have electrodes implanted in their brains. The electrodes emit low-voltage electric pulses that somehow interfere with the electrochemical signals of the brain's neural networks and make the symptoms of this terrible disease go away. DBS has been successfully applied to other neurodegenerative disorders such as Alzheimer's and Tourette's syndrome. The physiological mechanism of DBS is not fully understood, but this is typical of complex cybernetic systems. Feedback loops evoke *emergent properties* in cybernetic systems that are often impossible to understand by reducing the system to its constituent parts.

Along with a host of other medical technologies, these medical cybernetic prostheses may be considered restorative because they serve the virtuous goal of restoring health and provide patients with quality of life. They form part of the technological evolution of medical devices that include not only mechanical prostheses but also new drugs.[16]

Nevertheless, there is another category of cybernetic prostheses that has been gaining increasing commercial significance in recent years, but which has nothing to do with restoring health. These prostheses aim to enhance. The notion of 'enhancement' has come to mean modulating behaviour, in other words doing things 'better'. The feedback loop of the cybernetic system uses the brain as the central processing unit that takes the signal from the prostheses, analyses it and decides a preferred course of action. The prosthesis then informs the brain how successful the decided action was, and the loop is repeated.

Examples of cybernetic enhancement include nootropic or 'smart' drugs such as methylphenidate, which is used by college students to help them study longer hours. More recently, a host of 'wearable computing' products has hit the market. Nike's Fuel Band helps users to monitor and stimulate their body's activity, and thus become more active and healthy. In 2013, Google launched 'Google Glass' which enhances human communication capabilities by connecting the wearer constantly to the Internet. Google Glass takes our contemporary symbiosis with social networks to the next level; the wearer is always connected, he or she becoming an information node in the global telecommunications network. In 2002, British scientist Kevin Warwick, presaging Google Glass somewhat gruesomely, had a hundred electrodes surgically implanted in the median nerve fibres of his left arm. Using the electrodes, he connected his nervous system to the Internet and thereby controlled a host of electrical devices including a robotic arm, a loudspeaker and an amplifier. Meanwhile, a global movement calling itself 'Quantified Self' promotes the idea of measuring everything about our body, behaviour and outcomes, and use these measurements as feedback signals for self-improvement. In 2014, Apple unveiled the Watch, which is likely to evolve into a wearable device supporting a quantified lifestyle. Could all such innovations be harbingers of our next evolutionary step? Are we destined to fuse with our computing machines and become absorbed into a collective superorganism of information, like the Borg's Hive? Perhaps Google would like that. They own a copy of the whole of the Internet. If we become cyborgs Google could become 'the rulers of the world'!

Enhancing cybernetic prostheses are based on technologies, but they are, essentially, cultural metaphors. They are part of the contemporary narrative of the brain (or the self) as computer, which increasingly means being connected directly to the digital world. Yet these metaphors are very powerful and we should take them very seriously indeed. They point to a new value system that affects society, politics and how we understand our future and ourselves. Thanks to new composite materials, miniaturisation and digital technologies, mechanics have become ultra-high-tech and the object of desire. More than that, they have acquired imaginary powers. Movies such as *Iron Man* promote the fiction that mortals can become god-like superheroes given the right technology. Self-enhancement is thus portrayed as a process of gradually 'upgrading' one's body to a mechanical one. Sometimes the transformation from human to super-human occurs by donning a special suit like industrialist Tony Stark does in *Iron Man*. Equally often the transformation happens in an immaterial void: Neo, Trinity and Morpheus in *The Matrix* are also cyborgs; they connect to the Matrix via cables plugged into holes at the back of their heads. After they do so their bodies are 'digitised' and acquire superhuman powers.

Exchanging one's weak, disease-prone biological body for an enhanced one, digital or otherwise, sounds like a good deal. In a consumerist age, in which the value of human exchanges is frequently monetised, futurist propaganda promises disembodiment as the way forward. Fashion models thin as sticks parade on global catwalks under the techno-thumbing of Kraftwerk's 'The Robots'. Perhaps anorexia is a 'female' neurotic reaction to this imminent cybernetic feature. The corresponding 'masculine' neurosis involves going to the gym in order to look like Arnold Schwarzenegger in *Terminator*. In both male and female neuroses, the body transforms into a sheath that covers the 'new' essence of a personhood that is essentially digital. Expressionless faces in fashion magazines point to hedonism without feelings, sex without emotions, to a cybernetic transaction between bodies not unlike the transaction between computer servers exchanging coded 'yeses' and 'nos'. Humanness is redefined as a chimera between a body we can ditch and a mind we can upload. Dualism has returned with a vengeance, only now the separation is more Platonic then ever; it is between form and substance. The outside form is the new perfection achievable by a man or for a woman, insanely muscular or anorexic; the inside substance is immaterial.

The many nuances of personhood have been simplified and replaced by profiles on social networks. As in the *Six Million Dollar Man*, the human body is assigned a monetary value. Increasingly, the body is traded and marketed; patented genes, egg and sperm banks, prostheses, personalised drugs, and an illegal and vicious organ transplant market are testament of this. Governments encourage the sick to 'shop around' for best prices for medical treatment in health 'markets'. Banks and insurance companies store information of our 'net worth'. Marketing executives talk about the 'lifetime value' of a customer.

These ideas of cybernetic disembodiment feed into the contemporary narrative of advertising and media. They may not be discussed as explicitly as I have done here, but implicitly they are everywhere one looks. They inform fashion, military tactics, government policies, corporate cultures, video games and business models on the web. At the same time, social networks erode previous social structures and reintroduce tribalism into our post-industrial societies. Marketers nowadays talk about 'tribes on the Internet' when they plan the next marketing strategy using social media. Consumerism is transforming and adapting to this new paradigm, which is reminiscent of pre-industrial epochs. We are all different now and we all have individualised – or tribalised – needs. Accordingly, businesses are shifting from manufacturing massively replicated products and providing services distributed through traditional retail channels to producing personalised products and services distributed directly to customers connected on the web. The distant future looks like a revision of the distant past.

As if we have come full circle since the big bang of the Palaeolithic, cyborgs are the new shamans. When we wear Nike's Fuel Band or Google's Glass, or any new enhancing cybertechnology that may come along, we are 'painting our bodies' with the symbols of a new totemism. The iconic half-man half-lion of the Stadel Cave reincarnates in the twenty-first century as half-man half-machine. Viewed this way, cyborgs communicate with the new, unseen gods of Artificial Intelligence. For, if we continue to replace body parts with mechanical prostheses one after another, we will arrive at an all-mechanical being, an intelligent robot. The logical conclusion of cyborg reduction is to become non-human. In this new totemism, the intelligent non-human has enhanced capabilities, a stronger body, omnipresence and immortality. It has all the characteristics of the old gods – and with the added benefit that we can actually create these gods in our factories and research labs. They are material gods in our own image. When we imagine ourselves as cyborgs we imagine ourselves uniting with these new digital gods of infinite wisdom and intelligence. Ironically, the only thing these new gods ask of us – not unlike the old ones – is our soul. To unite with them we must surrender our humanness.

The cyborg is the metaphor for life and personhood in the twenty-first century. But unlike previous metaphors, like mud or hydraulics, the cyborg is a metaphor not of the present but of the future. In typical postmodernist style, the cyborg has to some extent undermined Artificial Intelligence before Artificial Intelligence has had a chance to really happen. Ironically, once again we worship gods that do not exist.

The fifth element

Cyborgs, like all cultural phenomena, are the products of our cognitive system. We saw how the evolution of language in the *Homo* lineage, from strictly social to general-purpose, enabled new cognitive abilities to develop in our species around about 100,000 years ago. As a result, our Palaeolithic ancestors started burying their dead and creating art. Art expresses symbolic reasoning imbued with theory of mind. When we create a piece of art we communicate a message to other members of our tribe, implicitly assuming that they have a mind like ours, capable of deciphering and understanding our message. Symbolic reasoning furnished our cognitive systems with new abilities, creating a new awareness where the external and the internal blended. Reality became a construct of language. Art objects were alive like everything else, social agents interwoven in the extended social fabric of our prehistoric ancestors, which included animals, trees, rocks, everything. The creation of Baudrillardian simulacra and simulations was the first thing that the modern mind did as soon as it was born. It did so through anthropomorphising, storytelling, an innate belief in dualism, and the use of metaphor. These are the four, uniquely human, aspects of our mind that have shaped what we invent, debate and often die for, since our earliest beginnings. They continue to shape the values, hopes and nightmares of our contemporary, highly interconnected society of globalisation and computer technology.

Understanding these four aspects is crucial for any discussion about the nature of the mind and whether it can be recreated artificially. That is why, before examining the philosophy, science and technology of the mind, it was so vital to get a clear perspective on the power of metaphor. But are we truly metaphor's eternal slaves, as the post-structuralists suggest? Isn't there an escape from the metaphor? Are we programmed by our nature to forever delude ourselves and never truly comprehend reality?

Thankfully, evolution has fashioned one more by-product of accidental neural rewiring that has had improbable implications for what we are and for the world we are creating. This 'fifth element' differs from the other four mental aspects because it has the potential to go beyond the ephemeral or the personal, and to ponder the abstract and the absolute. Indeed, it has invented these terms: 'abstract' and 'absolute'. It is the deepest and most mysterious aspect of our consciousness: it is our self-awareness. The ability to think that you think, to observe one's thoughts, to be aware of oneself and one's inner world, to have an inner conversation, is the escape hatch from the limitations of our evolutionary history. Self-awareness is the route to salvation from our metaphors and delusions. It is what gives us the gift of free will. For although we are descended from a long line of apes we mutated so significantly that we acquired the astonishing capability to look at ourselves from a standpoint outside ourselves. In doing so, we can see our limitations and do something about them.

But how can this be? Where are 'we' when we look at 'us'? How can the subject become the object, and vice versa? How can we be in two different places and still be one and the same? Self-awareness may be our salvation but it is also the ticking bomb that lies at the foundations of logic, mathematics, computer science and Artificial Intelligence. It suggests that the mind is capable of observing itself. This is the paradox of paradoxes, equivalent to suggesting that an eye can see itself, or a hand can hold itself.

All these statements are completely illogical, and that is why the existence of self-awareness has baffled logicians, philosophers, scientists and mathematicians since ancient times. And yet *we know* we have self-awareness. Perhaps, as Descartes said, it is the only thing we can be certain of.

That, however, does not mean that we also understand the mechanism of self-awareness. There seems to be a feedback loop in our minds that constantly feeds information about our self. At any time we can switch our self-awareness on and become aware of what we do, where we are, what we think. The big question is to whom does self-awareness feed this information? If we are the story, who is the narrator? Where is the 'I'?

Feedback loops are prevalent in control theory and cybernetics, as well as in every living system. Could self-referential functions where external signals are amplified and fed back into the system be the quintessence of the self, of intelligence and consciousness? Many have been led to the logical conclusion that 'we' *are* the feedback loop,[17] and that consciousness is nothing but the recursive processes of feeding information through the neural pathways in our brain. *Are* we feedback loops? Or is there something beyond the physical brain that causes self-awareness? Is the soul made of software?

Equipped with an understanding of the evolution of the mind and therefore its limitations, let us now venture onwards. While remaining vigilant and aware of the impact and influence of cultural metaphors, let us consider self-awareness and two further areas inextricable from the human quest for knowledge: philosophy and technology. Let us see how these two disciplines struggle to master the four basic aspects of the human mind by deploying a fifth element – self-reflection – and how this fifth element has come to haunt Artificial Intelligence.

PART II

THE MIND PROBLEM

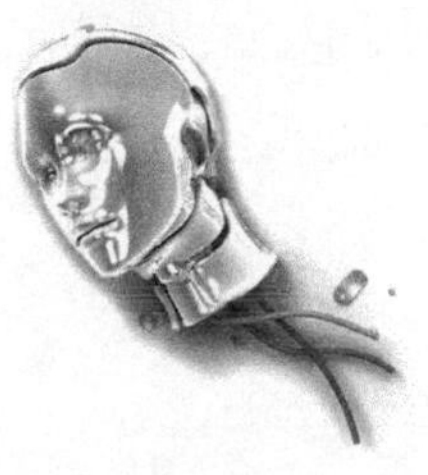

What we cannot speak about
we must pass over in silence.

Ludwig Wittgenstein

What is the mind?

When we aim to program a machine to 'think', what exactly do we mean? What exactly *is* 'thinking'? Who, or what, does the thinking in us humans? Is it our brain – that wet, convoluted, mushy, material object? People who are religious might argue that thinking, feeling and self-awareness lack physical substance, and that consciousness is the manifestation of an immaterial soul. After all, we often 'experience' the 'I' of our mind as existing outside our body. Just think of dreams, or altered states of consciousness, or simply reading a novel and becoming transposed to a totally fictitious place and time. Most of us remember experiences where our body seemed absent. You may be somebody who holds a belief in an immaterial mind such as this. This does not necessarily mean you are religious. There are scores of non-religious scientists who think similarly, although few would admit to it. Often they will use a vocabulary different from a metaphysical or religious one, in order to express very similar beliefs. For instance, prominent and self-declared agnostics, including the physicist Stephen Hawking, proclaim that human consciousness resembles a software program, and that at some time in the future it will be possible to extract it from your biological body, download it on a computer, so 'you' may live digitally forevermore. But isn't this just saying in other words what religious people have been preaching all along? That body and soul are two separate entities uniting at birth and separating at death? How come 'software' sounds so much like a modern synonym for 'soul'? And what about the 'computer'? Could it be the twenty-first-century synonym for 'heaven'?

Others scientists, such as the celebrated mathematician and inventor Stephen Wolfram, go a step further. They claim that the *whole* of the universe is a computer program. Our minds must therefore also be a piece of cosmic software, a program. We are part of the cosmic whole in the same way that specific-purpose programs integrate into bigger and more powerful systems to make up a supersystem. Many mathematicians and physicists share views such as these that resemble panpsychism, the metaphysical notion that everything has a spirit. And it is not a coincidence that mathematicians are relatively more prone to such views, compared with other scientists, or indeed engineers. For some weird and as yet unknown reason, the laws of nature can be expressed mathematically. Indeed, it seems that mathematics could be the essence of the whole universe. Take for instance the most widely accepted theory in contemporary physics, which is purely mathematical and is called string theory. It suggests that the fundamental particles that make up the material universe are created due to the twisted geometry of time-space. In other words, matter results from immaterial geometry. So is the universe essentially mathematical? Could there be two natures in reality – a materialistic one and a non-materialistic one – just like the dualist Descartes suggested? And, if so, which one has prevalence? Do numbers exist before we count something? The British philosopher and mathematician Bertrand Russell certainly thought as much, and he was not the only one. But, if so, where is the abode of numbers? Is there another reality beyond the one we perceive with our senses?

Diametrically opposed to the mathematical dematerialisation view of the cosmos and of the mind sit the doubting Thomases who believe exclusively in a purely materialistic world. The mind, they claim, is a biological phenomenon; it is what living, vigilant brains housed inside craniums create. Nothing else exists beyond what we can observe with our senses and our scientific instruments. These people are called materialist monists.

Idealist monists believe the exact opposite: that the material world is an illusion; only minds are real, they say, because everything that we can possibly know about the world is filtered through our minds. To substantiate their views they point, for example, to the fact that there are no 'colours' in nature; and that only our minds see 'red', or 'blue'. Isn't that evidence that we live in a world of pure ideas, rather than of molecules and atoms?

This tug-of-war between materialists and idealists, monists and dualists has persisted for centuries. It ties many up in knots of confusion, but also acts as a generator of great ideas and inventions. Our science, our fiery debates about how we should govern our societies, about what is ethically right and what is abhorrent, are guided by this quintessentially Western tension between two opposing views of the world: one firmly rooted in the belief that the material world is the product of ideas; the other purporting that the material world is all that there is. These two opposing views have been guiding and rejuvenating computer science and Artificial Intelligence since their beginnings in the late 1940s. They are also the source of doubts and disbelief about the promises that Artificial Intelligence makes. If the mind is immaterial, then how can we ever hope to construct a material computer with a soul? How can we force mindless electrons inside computer chips to become self-aware? Unless the human mind itself is a software program, in which case creating an Artificial Intelligence ought to be straightforward: we simply need to write the right program, and the program will think. But if we accept this proposition, we must ask ourselves who wrote *our* program? Are we trapped by the contemporary literary metaphor for life? Or is there something beyond the metaphor, a deeper insight into the nature and cause of being and becoming?

Ever since British mathematician Alan Turing wrote his seminal paper on machines imitating humans, various camps in computer science, robotics and Artificial Intelligence have been demarcated by the dichotomy between materialism and idealism. We cannot possibly gain insight into Artificial Intelligence, and its potential to change our world and our civilisation, unless we understand the centrifugal ideas that dominate it. To decide what to trust and what to reject, we must begin with the foundations of Western philosophical thought, and follow them all the way to today's ferocious battles of ideas about the mind. Time to board our imaginary time machine, turn its clock back around twenty-five centuries, and take a trip to Athens . . . .

7
A BLUEPRINT FOR A UNIVERSE

Plato was twenty-three years old in 404 BC when Athens lost the long war to Sparta. It was the last chapter of a disastrous struggle for supremacy that had lasted for almost three decades. The victors placed a garrison on the Acropolis, then quashed Athenian democracy and replaced it with an oligarchy of thirty select aristocrats. Plato was amongst those who were pleased with the new political arrangements. Two of his uncles from his mother's side were members of the new government. As an aristocrat, he despised direct democracy where the majority – the 'uneducated rabble' as aristocrats would call their poorer fellow citizens – ruled. In fact, he considered hoi polloi as the cause for the defeat of Athens. If it weren't for amoral demagogues and emotional swings in public opinion the city's generals and admirals would have made short work of the Spartans. So he welcomed the 'Thirty Tyrants' as the bitter poison that was necessary for the return of order and sobriety to his ruined and demoralised city. But it was not long until he became disillusioned. The Thirty Tyrants set off a wave of brutal persecutions against their political enemies. Hundreds were summarily executed, and thousands exiled or thrown in jail. When the dictators run out of public funds they turned to confiscating property from fellow aristocrats. That was the tipping point in their reign of terror. Within a year they were ousted by a counter-revolution, and democracy returned to Athens with the concession of the Spartans.

Like many young men and women today, Plato was an idealist who dreamed of a better world. But Plato was also destined to become one of the greatest philosophers of all time. So instead of a *better* world, young Plato started thinking about a *perfect* world. He became obsessed by a burning question: if the rule of the few and the rule of the many were both abject failures, how should people govern themselves? For Plato there *had* to be an ideal form of government that ensured order and prosperity for all. So he began to ponder the concept of a perfect *polis*, laying the foundations of all Western political philosophy that followed.

Plato came from a well-off family, and that meant he was given an excellent education. Like his peers, he had studied music and mathematics, and was deeply influenced by a mystical school of thought that was prevalent in fifth-century BC Greece, the 'Pythagoreans'. Founded by Pythagoras the mathematician, adherents of the school believed that numbers constituted the true nature of things. They also believed that the world was infinite, and in the transmigration of souls. Their ideas would find a place in Plato's philosophy, and become distilled by his genius into a sophisticated system of thought that has lain at the foundations of Western civilisation ever since.

Plato not only happened to be born at a crucial time historically for Europe but was also fortunate to have had Socrates as his teacher. Indeed, it is Socrates' voice that 'speaks' in Plato's works, all of them written as dialogues where Plato's teacher unravels the misconceptions of fellow interlocutors. This 'unravelling' is central to Socratic, and Platonic, thinking. It implies that we are deluded by virtue of being human. So whatever we think we know is wrong. This concept of inherent self-delusion forms the bedrock of epistemology, the philosophical method that explores how we know things, and how much we can possibly know about anything. Socrates made it his life's goal to show how mistaken we are when we claim to know something. He would go to the Athens market and start a conversation, or a dialogue – hence the word 'dialectic' – with a fellow Athenian. After a while Socrates' interlocutors would become very frustrated with his persistent questioning and doubting, but, as the dialogue went on, they would finally arrive at a point where they had to admit their total ignorance. That point – called *aporia* in Greek – was according to Socrates the first step towards wisdom, for wisdom began with the realisation of ignorance. His famous words were: 'The only thing I know is that I know nothing.'

The restoration of democracy in Athens in 403 BC brought a fresh wave of persecutions of which Socrates, a critic of democracy and a supporter of Sparta, became a victim. Although the formal accusations against him were that he was irreverent to the city's gods, it seems that the real motive for Socrates' persecution was to hurt his aristocratic friends who had sided, at least initially, with the Thirty Tyrants. He was put on trial and sentenced to death by drinking a mixture containing the poison hemlock. It was a terrible – and unjustified – end to one of history's greatest minds. Plato left Athens soon afterwards and travelled to Sicily, Egypt and Libya. He became acquainted with, and was thoroughly impressed and influenced by, the Egyptian civilisation. He began to write incessantly, worked as the political guru of tyrants,[1] was repeatedly thrown into prison for allying with the wrong people, and ended up being sold as a slave – to be later freed by an admirer. At the age of forty he returned to Athens and became the founder of the Academy.[2] By then he was convinced that he had cracked the problem of perfect government, and much more.

Shadows in a cave

In Plato's *The Republic* the character of Socrates not only explains the principles of a perfect government but also relates them to the human body and soul. Society, he suggests, should be organised in three castes. The lowest caste should consist of the productive people, such as labourers, farmers, merchants and artisans. They represented the abdomen of society and corresponded to the appetite part of the soul.

The middle class should be made of the guardians, the strong and brave members of the armed forces. They represented the chest of society and corresponded to the spirit part of the soul. The head – or the reason part of the soul – should be the rulers, who must be intelligent and rational, in love with wisdom,[3] and therefore the best at taking decisions for the community. They would be the 'philosopher-kings'. In Plato's own words: 'Until philosophers rule as kings or those who are now called kings and leading men genuinely and adequately philosophise, that is, until political power and philosophy entirely coincide, while the many natures who at present pursue either one exclusively are forcibly prevented from doing so, cities will have no rest from evils, . . . nor, I think, will the human race.'[4]

Plato's politics have had a tremendous influence in Western political thinking. They seem to uphold a reasonable tenet: that the affairs and decisions of government should be reserved for those who are morally and intellectually superior. The word 'philosopher' means 'one in love with wisdom'; in today's context this would include scientists, engineers, technocrats, and virtually anyone with specific knowledge of something. Shouldn't we trust the running of our government to those with the appropriate skills and moral integrity to act impartially, unselfishly and for the benefit of all?

Not everyone is convinced. The German philosopher Karl Popper has argued that Plato's idea of philosopher-kings lies at the root of totalitarianism.[5] Take, for example, communism, which is based on the dictatorship of the proletariat, or fascism, or Nazism: these ideologies purport that society should be run by an enlightened elite. Grounded on the arguments in Plato's *The Republic*, totalitarians claim intellectual and moral superiority, and therefore the right to rule. In the eyes of those who believe in liberal democracy, no human is either infallible or morally superior to others. Besides, totalitarian ideologies have failed, if the twentieth century has anything to teach us.

But what if those philosopher-kings were not fallible humans, but infallible intelligent machines? What if their morality was neutral, and their intellectual ability several orders of magnitude higher than any human? Shouldn't we let those benevolent, hyper-intelligent machines rule our world?

Some, of course, would rush to answer that question with an emphatic 'no'. The idea of machines, benevolent or otherwise, ruling over human affairs seems like a non-starter. Nonetheless, the question ought to give us pause. Historically, all totalitarian ideas were populist; they promised riddance of a corrupt, ruling elite. In practice, they ended up replacing one corrupt system of government with another, also corrupt but more brutal. But totalitarianism led by intelligent machines could offer a new social contract: to be ruled purely by perfect reason and incorruptible goodwill. Given this choice, would you still prefer to be governed by corrupt politicians who collude with powerful industrial lobbies and make a mockery of our so-called democracy? Wouldn't you choose reason and wisdom over persuasion and rhetoric? Given the increasing alienation of many voters from the established political class, this is a proposition that might appeal to a great number of citizens. For better or for worse, Plato's politics are going to stay with us for much longer than Popper might have hoped. Later in this book, I will examine in more detail the ramifications of Plato's political views in a future world ruled by AI. But now let us return to *The Republic*, for therein lies another idea that is central to any discussion about the human mind and how it perceives the world, as well as whether it could be duplicated in a machine. Socrates articulates this big idea via the famous 'allegory of the cave'.

Socrates imagines us all living inside a cave as prisoners, with our heads and legs restrained, able to stare only in one direction: towards the wall, at a procession of shadows. Not knowing otherwise, we think that the world *is* the shadows. But in reality we are prisoners of ignorance. Occasionally, someone manages to unshackle themselves and escape from the cave, creep outside into the open and stare at the sun. This individual gradually comes to understand what the 'real' world is like. According to Socrates, this person, the enlightened one, has a moral obligation to return to the cave in order to set the rest free as well. Unfortunately, he adds, it takes a lot of convincing to make people abandon the chains of their delusions; which also explains why Athenians were so irritated and annoyed with Socrates.

With the allegory of the cave Plato introduces his theory of forms, the most fundamental concept of his philosophy. The theory of forms stipulates that the world we perceive with our senses is not the reality but an *imperfect projection* of reality – a shadow puppet show on a cave wall. 'True' reality is made of *perfect forms* that exist in the realm of ideas. Thus, there are two worlds: the apparent world and the unseen world of ideal forms. Plato's philosophy is grounded on the assertion that the ideal world is the *cause* of the apparent. In effect, Plato adopts and expands the mathematical and metaphysical ideas of the Pythagoreans. For instance, in *Timaeus*, another of his books, Plato associates the four classical elements that the Greeks believed the universe to be made of (earth, air, water and fire) with geometrical solids.[6] Earth is associated with a cube, air with an octahedron, water with an icosahedron and fire with a tetrahedron. Ideal forms were mathematical in nature. For Plato, mathematics *creates* the world. The abstract, ideal, perfect, symmetrical solids are the *causes* of the elements of earth, air, water and fire. We walk, breathe, quench our thirst and warm our bodies with illusions created by a deeper, geometrical reality.

Plato's theory of forms is often regarded as 'mystical', and there are several good reasons for this. He never explains what the world of ideas is made of. Instead, he suggests that knowledge of this world is possible via reason and is experienced as remembrance. The real world 'reveals' itself to those who strive to 'remember'. Echoing the meditative practices of the Pythagoreans, Plato bequeathed a formal method of justified knowledge by means of pure reason. If something could be shown to be logical then it should exist. This is an extraordinary argument. It suggests that if you can conceive a perfect idea then that idea must exist in reality. Indeed, it *is* reality. Logic and ideas thus take precedence over experience. For Plato, the water you drink is not really water; it's the projection of solids with twenty faces – and that's because Plato came with a perfectly good logical argument in support of this hypothesis.

Plato's methods and ideas were picked up by philosophers of the Hellenistic and Roman times, and found their way into Christianity. The triadic God defined by the First Christian Council of Nicaea in AD 325 is modelled after the Neoplatonic, tripartite concept of the universe. The Father-Creator corresponds to the productive class of Plato's *The Republic* and is the 'belly soul' of the world. The Holy Spirit is the Guardian, and the Son the all-ruling and all-judging Reason, or 'Logos', the Word. As Christian theology developed further over the ensuing centuries Platonic ideas became more prevalent, especially in the East. The Bogomils, a heresy that gained dominance in eleventh-century Byzantium,[7] believed that the material world was an illusion. In the fourteenth century, the Greek theologian-monk St Gregory Palamas (1296–1359) founded quietism (*hesychasm* in Greek) based on Platonic ideas, the monastic tradition of eastern orthodoxy still prevalent in the twenty-first century. On the verdant slopes of Mount Athos in northern Greece today, orthodox monks practise rhythmic breathing exercises and the repetition of a mantra-like prayer in order to 'ascend' from the dark cave of ignorance to the eternal light of sun-like Jesus.[8] For them, the experience of God comes through prayer and meditation, by training the mind. Union with the eternal is achieved by shutting off the external world of senses.

In the West, Platonism remained the dominant idea in the Catholic Church for many centuries, mostly thanks to St Augustine (AD 354–430), who, before becoming a Christian bishop, was a devout Neoplatonist. However, Platonism was ultimately dethroned in the late thirteenth century by St Thomas Aquinas (1225–1274), who pointed to Aristotle as his main source of inspiration. Aristotle was a student of Plato, and a typical case of the student surpassing his master. Contrary to Plato's mysticism, Aristotle suggested that it was only possible to discover the world of forms through investigation of the natural world. As we shall see, Aristotle provided the foundations of Western empiricism, the polar opposite of mysticism. By adopting an Aristotelian worldview for the Church, St Thomas Aquinas sanctified, and by consequence enabled, scientific enquiry. It is thanks to him that Western Europe managed to compete effectively with the technologically advanced – and very Aristotelian (until the late Middle Ages) – Arabs. When Islam reverted to mysticism and authoritarianism, and thus waned as a world power, Western Europe had already adopted empiricism at its core. Galileo could embark on the discovery of reality by peering at the heavens through a telescope, rather than gazing at his navel like the orthodox monks, or dancing in circles like the dervishes of Istanbul.

The laws of nature

With Galileo Galilei (1564–1642) in Italy and Francis Bacon (1561–1626) in England, Europe was ushered into the glorious era of scientific discovery by the late sixteenth century. Empiricism and the scientific method took over from mysticism. It was a historical watershed without precedence. Our species has been making observations of the natural world since prehistoric times. Effects were linked to causes. Causality in nature was evident, but not understood. Well into the early seventeenth century, natural phenomena were still attributed by many thinking people to supernatural causes. Divine providence was assumed to pull invisible strings behind every manifestation of nature. The scientific method provided an alternative, revolutionary way of understanding causality in nature. Instead of simply believing, one was now compelled to justify that belief by experimentation. But what exactly *is* experimentation?

Natural phenomena have regularities. You light a fire under a pot full of water and a little later the water begins to boil – *always*. Experimentation is grounded on the principle of regularity: given the same cause the experimenter should expect the same result. It follows that, by altering incrementally the degree of the cause, one could observe and measure incremental changes in the result. In the Renaissance,[9] new scientific instruments were manufactured that allowed the results of experiments to be measured. Accurate instrumentation became the enabler of scientific progress, because naturalists understood that by measuring something they could begin to understand it better. It is almost impossible for us today to comprehend the level of amazement that these early scientists must have felt when they started to investigate nature experimentally. Measurement assigned numbers to regularities, a transformation that offered new perspectives and insights. The early scientists could now manipulate the numbers in order to draw new conclusions about the relation between natural causes and effects. This mathematical transformation of observations led them to the discovery of the first physical laws that culminated with Isaac Newton's laws of gravity. It was an incredible journey for the European mind, set in motion by Aristotelian empiricism and arriving at something completely unexpected: at the curious, and disturbing, confirmation of Plato and the Pythagoreans! Scientific discoveries showed that numbers and mathematics ruled the universe. As foretold in *The Republic*, scientists (or 'natural philosophers', as they used to be called) had left the cave of shadows and now stared at a mathematical sun. Amazingly, Plato had made a triumphant comeback to the forefront of European philosophy through the back door of physical mathematics.

The Ottomans are partly to 'blame' for this. Their capture of Constantinople in 1453 caused many Greek scholars to emigrate to Italy, bringing with them precious manuscripts of Plato's works. The chief influencer amongst them was the Neoplatonist philosopher Pletho Gemistus (1355–1454). Pletho was a teacher of the last Greek imperial dynasty of Byzantium who struggled to unite the Eastern and Western Churches and instigate a European crusade to defend Constantinople. He failed, but in the process he made powerful friends in Italy. The Medici, the ruling family of Florence, built a Platonic Academy dedicated to Pletho. They also commissioned translations of Plato into Latin. Until then Plato's works had been virtually forgotten and were mostly unintelligible to Western Europeans. Catholic clerics considered Plato very 'Greek', which was a euphemism for being a heretic.[10] Greek language was scarcely studied at European universities and the Greek manuscripts that circulated were translations from the Arabic. The Medici ensured that Plato was read once more by the brightest European minds. The reintroduction of Plato in Europe in the late fifteenth century and the apparent affirmation of the Platonic theory of ideal forms by Aristotelian science created a unique cultural tension which is distinctly Western European and remains to this date. At the centre of this tension lies a dichotomy between form and matter. Which one takes precedence? Is matter the cause of form? Or is form the cause of matter? Western civilisation seems as yet unable to decide.

Progress in science since the scientific revolution has left this cultural dichotomy unresolved. Our modern understanding of the cosmos is based on reducing matter to elemental building blocks, a process called 'reductionism'. Science has triumphed in the past five centuries because it has managed to explain most natural phenomena by means of observation, experimentation and measurement. It has 'reduced' the apparent complexity of the cosmos to a few simple laws that hold true across the whole of the universe. It has given us concise theories of how this universe may have come about, and how life may have evolved.

There are, however, still several gaping holes[11] in this truly magnificent edifice of knowledge. Take, for example, the Standard Model of physics, which explains how matter is structured by means of elementary particles called 'quanta', and how the forces of nature[12] also act by means of quantum particles. The Standard Model is undoubtedly one of the wonders of modern science. However, it does not yet explain how gravity acts at the microscopic scale of elementary particles. Gravity refuses to be reduced to a fundamental particle, to a 'graviton' that would mediate gravity at the quantum level. 'Quantum gravity' remains one of the gaping holes in modern science. Several theories have attempted to fill that hole, but the one that is most widely accepted by physicists is string theory. String theory is a purely mathematical-geometrical theory that stipulates certain mathematical entities *as the causes* of the elementary particles scientists have observed and classified in the Standard Model. Plato could not have hoped for a better vindication. As we have seen, he stipulated[13] that a specific number of symmetrical solids – i.e. pure geometrical shapes – gave rise to the fundamental 'elements' of matter, and therefore to the whole material universe. String theory suggests a very similar causation. It is therefore not surprising that many scientists are essentially Platonists, particularly the ones working in mathematics, physics, cybernetics and computer science.

The anthropologist Stefan Helmreich has studied how scientists and technologists think in Platonic terms, by spending several months at the Santa Fe Institute.[14] The Institute is renowned for its pioneering research into chaos theory, cybernetics and complex systems. One of the research areas in the Institute is the development of simulations of biological systems in a computer – what we might call digital 'Artificial Life'. Helmreich's interviews with scientists working on digital Artificial Life revealed that most of them believed that the world was essentially mathematical. Their computer simulations of biological systems showed that complex biological behaviour could be recreated by a few simple mathematical rules coded in a computer. By recording their scientific beliefs, Helmreich's study revealed how the form/matter dichotomy at the core of Western thought since the Renaissance is nowadays articulated using terms borrowed from computer science.

Artificial Life in the Santa Fe Institute's labs seems to verify Plato's Theory of Forms: software is the *cause* of digital life. But could 'real life' also be the result of a 'code'? Could the whole universe be the result of information processing? There apparently seems to be a blueprint for the stars, the galaxies, the quanta, the Standard Model, for you and for me. That blueprint is called 'physical laws' and is expressed in pure mathematics. This logical conclusion, stemming from the mathematicalisation of nature as well as computer simulations of natural phenomena, has become very prevalent in the global scientific community and is central to the problem of Artificial Intelligence. Indeed, it has become pivotal to the problem of consciousness. For, if we assume that the material universe is created because of the underlying mathematics of string theory, and if we accept the logical inference that our minds are also created because of that same fundamental mathematics, then it can only follow that consciousness *is a fundamental property of the universe*. Does that mean that the universe possesses an intelligent, ideal mind? That there exists a universal *nous*? Could this *nous* be the Supreme Software Designer of our material universe? Was Plato right?

Platonic proteins in the brain

English mathematician Roger Penrose and American anaesthesiologist Stuart Hameroff seem to think so. In fact, they claim to have discovered the cause of consciousness and to have linked it to quantum computing.

It all started with Hameroff's research, which included the study of microtubules (MTs) – structures of proteins called tubulins that exist in our brain's neural networks. MTs are very important with respect to the shape and function of brain cells. They help neurons form connections with other neurons and they are probably implicated in learning and understanding. Hameroff noticed that during anaesthesia, microtubules in the brain changed shape and were 'deformed'. When proteins deform their properties change. This is a characteristic of protein function.[15] Hameroff concluded that there had to be a connection between the deformation of microtubules and the loss of consciousness by patients under anaesthesia. Further studies showed that the proteins that make up the microtubules deformed because of so-called 'London forces', named after the German-American physicist Fritz London. Basically, this meant that tubulins could deform in only two ways, which were determined by the position of an electron inside the protein. Because an electron is a quantum particle, modern quantum physics tell us that it can exist in many places at the same time, a state called quantum coherence. Thus the tubulin's state is a macroscopic projection of the electron's quantum state. This means that the tubulin can be in a superimposed state as well, a state when the electron/tubulin is still 'undecided' about which of the two deformations it will ultimately take. The tubulin exhibits all the characteristics of a quantum computer – where instead of *two* discrete binary states (either '1' or '0') as in classical computers there is *one* combined state (i.e. '1' and '0' at the same time).

At the time of his research Hameroff had not yet realised the connection of tubulins and microtubules to quantum computing, until he read Roger Penrose's book *The Emperor's New Mind*.[16] In that book, Penrose attacked Artificial Intelligence by suggesting that it would be impossible to code consciousness in a computer. He based his arguments on the limits of logic and suggested that because computers are programmed using logic, they would always be lacking in comparison with real human beings. Humans, he claimed, have capabilities beyond logic, such as intuition, which are incomputable. He concluded that consciousness must be a quantum phenomenon.

In collaboration with Hameroff, Penrose expanded his theory by showing how microtubules could be the seat of consciousness. In what is called the 'orchestrated objective reduction' (Orch-OR) model of consciousness, Penrose and Hameroff suggest that consciousness is embroidered in the fabric of reality.[17] According to Penrose, space–time is not continuous but granular. If we imagine a fantastical microscope that could peer into the most minute aspect of space–time, we would notice thick foam made of multi-dimensional geometrical shapes. Penrose calls this foam 'spin networks', and considers it the fundamental structure of reality. Mass and energy are manifestations of quantum phenomena at the level of these infinitesimal spin networks.[18] And so *is* consciousness: quantum fluctuations at the fundamental geometry of space–time cause electrons in the tubulins to 'quantum decohere', i.e. to choose between the two possible places that define which way the tubulin will deform, and thus switch consciousness on or off.

The quantum hypothesis for consciousness proposed by Penrose and Hameroff has received much criticism by physicists, neurobiologists and logicians. The problem with the hypothesis is that it is founded on two fundamental assumptions, both of which could be wrong. The first assumption is that, because logic has been proven mathematically to have limits,[19] these limits preclude the coding of conscious machines. The second assumption is that consciousness will never be explained as a purely biological phenomenon. Both these assumptions can be shown to be false, as I will explain later. Nevertheless, the core idea of explaining the mind with geometry retains many apologists, and not only amongst the multitudes charmed by the dual mysteries of quantum physics and consciousness without really understanding either. It connects a physical and observable phenomenon at the molecular level – the deformation of microtubules in neurons – with an observable phenomenon at the multicellular level of a patient who is unconscious under anaesthesia. And explains the molecular phenomenon with a deeper cause that could be theoretically simulated in the future, when we will hopefully have quantum computers. It is a potentially testable hypothesis. Perhaps, then, it is too early to dismiss the principles of the orchestrated objective reduction model of consciousness as nonsense. So instead of criticising it, let us examine its ramifications.

If Penrose and Hameroff are right and consciousness is a fundamental property of the universe at the quantum level, entwined with its geometry, then the current premise of Artificial Intelligence is false. There cannot be truly intelligent machines for it is impossible to code something – i.e. a mind – that exists at the ultimate causal level of reality. It would be a logical impossibility if we did: it would mean that we can reprogram the universe, and therefore ourselves, and that would imply that we exist outside our universe, which we do not and cannot.[20] However, if we accept this thesis then we must also accept that our minds are independent of our bodies. We must accept that our minds exist in the quantum geometry of spins and that our physical brains are mere receivers and not producers of self-awareness, like radios receiving a signal from another source that could be miles away. This is virtually identical to accepting the dualist position that there is an immaterial soul separate from the body. Indeed, the argument goes beyond that: it suggests that this immaterial 'soul' *is* in fact our mind, our consciousness, the true '*us*'. But could there be minds without bodies? Are we made out of pure information structured around mathematics?

8
MINDS WITHOUT BODIES

Our brain's neurophysiology generates experiences, such as dreams and altered states of consciousness, which are not of the material world. Such experiences can seem as 'real' as anything else. A famous quote by the Chinese philosopher Zhuangzi (369–286 BC) plays with this transgression between dreams and reality by juxtaposing the dream of a butterfly with that of a man. If you dream you are a butterfly and then wake up to find that you are a man, how do you know that you are not now the dream of a butterfly? In cases of lucid dreaming the dreamer is aware of dreaming, a fact that makes her experience all the more weird, wonderful or terrifying. Lucid dreaming feels like entering another dimension of existence. So could we be living inside somebody else's dream?

Our Palaeolithic ancestors dreamed of invisible worlds – as we do today – and used art to articulate their beliefs about them. Unlike us, however, for them all worlds, dreamed or undreamed, formed a continuum. Consciousness was not separated into dreaming and awakening. Whatever evidence we have from that era suggests that our distant forefathers and foremothers often came together to perform social rituals involving a combination of rhythmic dances, songs and the use of hallucinogenic drugs.[1] Led by shamans, they would enter special mental states similar to lucid dreaming. These invisible worlds were experienced 'outside' the physical body. The body was left 'behind' in the coarse realm of physical existence, like a simulation of death. These out-of-body experiences seemed to confirm that pure mind could exist by itself. Perhaps therein lie the ancestral roots of why, to this day, we instinctively differentiate the mind from the body.

But let's fast-forward to classical Greece, where many prehistoric totemic and transcendental rituals survived in the guise of new gods and narratives. Most prominent among them were the Eleusinian mysteries celebrated in honour of Persephone,[3] the goddess of the underworld. According to the myth, Persephone was the beautiful daughter of Demeter, the goddess of harvest and fertility. While she was out with friends picking flowers in a meadow, Hades, the god of death and the dead, fell in love with her, seized her and took her by force to his dark kingdom in the underworld. Demeter searched everywhere for her daughter, and when she finally found out what had happened she demanded of Zeus that Hades return her daughter immediately. In the beginning Zeus was not too keen to intervene. But Demeter found a way to twist his divine arm: she caused a terrible drought. Crops failed. Mortals had nothing to sacrifice to the gods any more. Zeus quickly found a compromise: Persephone would spend four months with her husband in the underworld and eight months with her mother in the world of light. And that is how we came to have the seasons, and why, during the four months of winter, nothing grows: Demeter is weeping for her beloved daughter. The mysteries were celebrated in late summer with a procession that began in Athens and ended in Eleusis, a city about twenty kilometres to the north-west. The initiates – including free men and women, as well as slaves – were then guided through successive experiences that evoked stages of the afterlife while remaining alive. Although only fragments of information exist about what went on during the initiations, it is very likely that a powerful psychoactive mix of opium and cannabis called *kykeon* mediated these out-of-body experiences.

One of the most illustrious initiates of the Eleusinian mysteries was Socrates. It is said that Socrates refused to enter the highest order of the Eleusinian priesthood because he would have had to take an oath prohibiting him from disclosing the shenanigans that took place in the inner sanctum. The great philosopher was adamant that he ought to share his knowledge with anyone who asked for it – so he quit. Nevertheless, the experience of the mysteries, in which the two worlds – the visible and the invisible – collided, influenced the fundamentals of Platonic philosophy. In Plato's *The Republic*, Socrates makes a distinction that would mark Western philosophy forever. First, he refers to the invisible world as *noeton*, a Greek word that means 'the one that is of the mind'. The visible, coarse world of matter he names *(h)oraton*, the one that can be 'seen'. He then suggests that the invisible world of pure mind is the most intelligible, while the external world of the senses the least knowable and the most obscure. According to Socrates you can only know what is in your mind or what you can reason about, and very little of what is 'out there'. For Socrates, as well as for Plato, the physical world is an illusion, a shadowy projection of ideal forms that can be only known by the mind. It would take another twenty-one centuries for this peculiar argument to resurface in Western thought. When it did – this time thanks to a Frenchman – it would form the foundation of the modern philosophy of mind and of epistemology – the philosophical method of knowing. As a result, the separation of *noeton* and *(h)oraton* – of body and mind, of software and hardware – would go on to bedevil the contemporary discourse on Artificial Intelligence and consciousness.

Cogito ergo sum

According to his own account, René Descartes became a philosopher thanks to a transcendental experience that he had on the night of 10 November 1619. He was garrisoned in Germany, in the city of Neuburg on the Danube, serving as a military officer under Duke Maximilian of Bavaria. To escape the wintry cold Descartes shut himself in a heated room (similar to a dry sauna) where, in the early hours of the morning, a divine spirit revealed to him a new philosophy. He emerged from the heated room a changed man, poised to shake the foundations of human thought. Some years later, in 1641, by which time he had already made major contributions in the fields of mathematics and geometry, Descartes published his seminal philosophical work, *Meditations on First Philosophy*.

The book is written as a series of meditations that last, like the biblical Creation, for six days. Following the apologetic methodology of medieval scholars, Descartes tries to prove the existence of God and of the soul by doubting the existence of the world. He thus distinguishes, like Socrates, two separate substances: the corporeal substance (which he terms *res extensa*) and the mental substance (*res cogitans*). The former is the essence of all physical things; the latter the essence of the mind, or of the soul, or consciousness. Descartes then arrives at two crucial conclusions. Firstly, that he cannot be certain that the external world actually exists: for all we know we may live inside the dream of a butterfly. Secondly, that he can be certain of one thing only: his mind and therefore his existence. By taking the Socratic argument to its logical conclusion Descartes suggested that we are, each and every one of us, the most important thing in God's creation. Why? Because we cannot be sure about anything else except ourselves. *Cogito ergo sum.*[3]

The repercussions of Descartes' meditations were enormous during his time, and their echo still resonates in the twenty-first century. With that single three-word sentence he laid the foundations of modernity, shifting the debate from 'what is true?' to 'how can we be certain about anything?'. The difference between these two questions has shaped our modern thinking and institutions, and here's how: 'Truth' requires an absolute authority: God, or his representatives on Earth. But 'certainty' is subjective and individualistic; you can only answer the question of how certain you are about something on your own and by yourself.

Before Descartes, the West perceived the world through the certainty of Holy Scripture, as a revelation of the truth directly from God. That certainty was now shattered beyond repair. Since Descartes, we have been living in an anthropocentric world in which the individual is the main actor, replacing the absolute authority of God and Holy Scripture. Thanks to Descartes' separation of *res extensa* and *res cogitans*, science was liberated from the shackles of the Church. Holy Scripture was not enough any more. Its statements about the birth of universe or mankind would now be seen as allegories at best; in terms of today's cultural relativism, they are regarded as one more creation myth amongst many. Scientists ventured forth to explore what the world was *really* made of and how everything came to be what it is.

Nevertheless, and for a very long time, scientists stayed clear of *res cogitans*. The things of the mind continued to fall under the jurisdiction of the Church. From the Enlightenment to the early twentieth century, science focused on the 'physical' world of coarse matter. Great breakthroughs in our knowledge were made in physics, chemistry, biology, geology and medicine by examining the material parts of the world, and of our bodies. Mental states, such as seeing red, desiring happiness, feeling pain, were considered fundamentally different from physical states and thus the phenomenon of the mind persistently remained beyond the scope of science. By separating the things of the mind from the things of matter, Cartesian dualism dictated the scientific agenda for more than two hundred years.

And yet there was an obvious question that begged an answered: how could non-physical states in the mind cause physical states in the body? What intervened between the two different worlds of soul and matter? If the mind and the body were made of different 'stuff', how did these two substances interact? This is the notorious 'body–mind problem', which for a very long time worried philosophers and theologians exclusively. It was only during the last decades of the twentieth century that science finally overcame its Cartesian dictate, turned its attention to the big elephant in the room, and began to explore consciousness as a material phenomenon. Yet, as soon as it did so, it crashed on to a huge philosophical wall built over many centuries with strong bricks of doubt.

As we shall see, contemporary neuroscience has achieved great insights into the human mind over the past twenty years. New examination methods and instruments – such as fMRI[4] – allow neuroscientists to peer into living, thinking brains, and record and analyse their thoughts. However, many contemporary philosophers doubt the potential of brain research to answer the important questions about the nature of consciousness. Their arguments apply to Artificial Intelligence as well, and the creation of consciousness in a machine. All arguments against AI derive from Descartes' body–mind problem and his introduction of subjectivity into how we are able to know anything about the world. So let us turn to these two issues and examine them more closely.

Monads, psychons and the pineal gland

Descartes was fully aware of the body–mind problem. He knew that he had to explain *how* the immaterial soul interacts with the material body. And here is the solution he suggested: first, he assumed that the body was purely mechanical, like a clock. It obeyed the laws of nature, gravity and the like. It was embedded in the physical world and consisted of physical matter. Nevertheless, emotions, thoughts and feelings were made of substances that did not obey the laws of nature. Instead, these phenomena were of the soul, which was housed inside the body and, in particular, inside the pineal gland. This is a small, nut-shaped gland wedged between the two hemispheres of the brain. Influenced by the medicine of Galen, that was still the orthodoxy in seventeenth-century Europe, Descartes observed that the pineal gland lay close to the ventricles that carried the cerebrospinal fluid. He posited that the ventricles acted through the nerves to control the body; and that was how thoughts were translated into actions.

The soul was the puppeteer and the body was the puppet. The only difference with this metaphor was that the puppet called 'body' had sensations. But how were sensations transmitted to the soul? And how was the soul informed about what happened in the external, material world?

To explain the soul–body interaction in the opposite direction, Descartes suggested that nerves were like a two-way highway delivering sensations to the pineal gland and causing it to vibrate in a sympathetic manner. A loud noise, the sight of a loved one, the smell of roses arrived at the pineal gland and stirred it as the wind stirs the surface of the sea. Those vibrations gave rise to emotions and thoughts: the loud noise was 'unpleasant', the face of the loved one gave 'joy', the smell of the roses was 'sweet'. The material world thus acquired emotional 'qualities'. The soul made sense of out everything. The soul was reason.

Descartes founded a new philosophical system in Europe that regarded reason as the chief source and test of knowledge. By instigating the separation of mind from matter, he claimed that truth was not sensory but intellectual and deductive. It was the mind that constructed reality by means of logic. Therefore reality had a logical structure. Physical evidence was unnecessary to ascertain the truth. Everything was in the mind, and reasoning was the ultimate method for arriving at the truth. This philosophical system is called *rationalism*.

The next great European philosopher to inherit Descartes' legacy and carry his theories forward was also a rationalist. His name was Gottfried Wilhelm Leibniz (1646–1716) and he was, arguably, the first person in history whose work combined consciousness with computers. Born in Leipzig at the end of the Thirty Years War that devastated the German-speaking countries, Leibniz became one of the foremost intellectuals of all time, making important contributions to just about every branch of science. He was also an engineer and inventor. He is considered to be the first computer scientist and information theorist. He advanced the binary numerical system that computers use today. He built calculating machines. Norbert Wiener, the father of cybernetics, claimed to have found in Leibniz's writings the first mention of the concept of feedback, the central idea of cybernetics. And, yes, Leibniz had a solution to the body–mind problem, too!

Not satisfied with Descartes' hypothesis concerning the pineal gland, Leibniz proposed the existence of elementary particles that could 'perceive' one another. He called them 'monads' and postulated that they were eternal and the ultimate elements of the universe.[5] Monads were immaterial and more fundamental than atoms. In a Platonic sense, they were ideal forms made of pure consciousness. The body–mind problem was therefore resolved as monads mediated between the non-material and the material realms. For Leibniz, consciousness was the most fundamental element of the universe; it was what the world was really made of. His notions predated the strikingly similar, contemporary ideas of Roger Penrose and Stuart Hameroff. These ideas are often called 'panpsychism' since they assume that the soul, or the mind, is a universal feature of all things, woven in the fabric of the cosmos.

Leibniz's form of panpsychism was to persist well into the twentieth century and resurface in the work of one of the most eminent neurophysiologists of our time, Sir John Eccles (1903–1997). Eccles was born in Melbourne where he studied medicine. Even as a student he was obsessed with finding an explanation for the interaction of mind and body, and this led to his becoming a neuroscientist. In the early 1950s, while working at the Australian National University, he began his research into the synapses of the peripheral nervous system. A synapse is a structure that permits a neuron to pass an electrochemical signal to another cell, neuron or other. Synapses are the connectors of the nervous system, and they also enable the nervous system to connect with the rest of the body. At the time that Eccles started his research no one knew how synapses actually functioned. By passing electric current along neurons and noting the results, Eccles discovered how the synapses acted as integrators that combined the excitation of many neurons and how, when this combination reached a certain level, they caused muscles to contract. By understanding the function of synapses, Eccles discovered how sensations and thoughts became actions in the body. It seemed as if he had resolved the body-mind problem that fascinated him for so many years! His work with neural synapses led him to sharing the 1963 Nobel Prize in Medicine.[6]

Following in the footsteps of Leibniz, Eccles – a devout Catholic – then went on to develop a philosophical theory that generalised on his discovery. He postulated the existence of 'psychons', mental units that represented a unitary conscious experience.[7] He then combined quantum physics and neuroscience. He linked psychons to bundles of neurons in the brain. When there was a thought or a will to act, psychons acted on the neurons and increased the probability of firing selected neurons through a phenomenon called 'quantum tunnelling'. During quantum tunnelling an elementary particle, for instance an electron, manages to burrow its way through dense, macroscopic classical matter, like a ghost going through a wall. This happens because quantum physics tells us that particles can also be waves. So the electron 'transforms' itself from a particle into a wave and wiggles its way through the empty spaces in the lattice of molecules. Quantum tunnelling is the mechanism upon which quantum computing is based. Eccles suggested that quantum tunnelling takes place in the neurons of the brain *because* of psychons. For him, dualism had to be accepted as a fact, because there was no other way to explain consciousness.

Nevertheless, not many agreed with Eccles because dualism poses a huge problem in science. It clashes with our fundamental understanding of the cosmos as expressed in physics, and with the law of energy conservation in particular. This law says that you cannot create energy out of nowhere, or out of nothing. Quantum tunnelling does not happen out of nowhere. Causality is preserved in quantum physics and energy is no exception. An electron burrows through a wall of matter because it is energetically excited by something in the material universe. If mind is immaterial (for instance, made of immaterial monads, or psychons), it could not affect something material like an electron. To do so the mind *would have* to be material.

This argument from the law of conservation of energy ought to suffice in order to silence all those who hold dualist beliefs about the mind. Yet apparently it does not. At the premier of a documentary about his life Stephen Hawking, one of the most prominent living physicists of our time, declared that the brain could exist outside the body.[8] He said, 'I think the brain is like a programme in the mind, which is like a computer, so it's theoretically possible to copy the brain on to a computer and so provide a form of life after death.' This is dualism, straight from the lips of a bona fide materialist. Hawking aims to circumvent the problem of conservation of energy in the body–mind problem by adopting the computer metaphor; while Eccles called quantum physics to the rescue.

But what is going on here? How can gifted physicists such as Hawking come out with such statements? How can a brilliant neuroscientist such as Eccles believe in immaterial psychons? Why can't they simply accept that the mind is a natural phenomenon made of matter, just like everything else? That our 'self' will ultimately perish – just like the rest of our body? What is so special about the mind that makes it hard to pin down and examine under a microscope, as one examines a neuron or a synapse? How come dualism persists in the study of consciousness? Why isn't everyone a 'physicalist'? Answers to all of the above seem to lie in the second big idea that Descartes introduced to the world with his famous three-word sentence *Cogito ergo sum*, 'I think therefore I am': the idea of subjectivity.

The hard problem

Imagine a world inhabited by zombies. Not like the wretches depicted in the movies. These zombies are special. They don't waddle and their limbs don't fall off when they walk into lampposts. In fact, they look exactly like ordinary human beings. And they behave like them, too. They have jobs, and families, and go on holidays. The only difference with 'real' human beings is that they *appear* to be thinking instead of *actually* thinking. They *seem* to be conscious, but they are not. You could prick them with a needle and they will shout 'ouch' – but in reality they would not feel any pain whatsoever. How could you tell the difference between them and a real human? What if you are the last real human person left in a planet full of these zombies? The simple answer is that you cannot possibly know,[9] at least not according to the contemporary adherents of dualism.

Thought experiments using these so-called 'philosophical zombies' (or 'p-zombies') aim to demonstrate that there is something about consciousness that cannot be verified objectively. This something is often called 'qualia'. The word comes from Latin and means 'what kind', or the qualitative aspect of something. In mind philosophy, the term is applied to describe the qualitative characteristics of the conscious experience. For example, the redness of red, or the bitterness of lemon, or the pain you feel when someone pricks you with a needle, your 'what-is-it-like' experience of the world. Qualia are used by modern dualists to argue against a purely materialistic – or 'physicalist' – interpretation of the mind.

The philosopher Thomas Nagel was perhaps the first to reframe dualism by using the 'what is it like to be?' argument for qualia. In his famous paper he asked, 'What is it like to be a bat?'[10] – a question that demonstrates the limits of science when it comes to exploring subjective experiences. It also represents an intellectual departure from our instinctive anthropomorphism. We are now wise enough to know that the minds of other animals are different from ours. So what is it like to be one of them? We cannot possibly know the answer to that – which seems to be a major problem in the physicalist approach to consciousness.

According to Australian philosopher David Chalmers, the existence of qualia creates an explanatory gap in our physical understanding of animal as well as human consciousness. For even if we ever do manage to understand every physical process in the human brain and explain memory, sensations, even thoughts, we will never be able to identify the physical processes of qualia. No one *else* will ever know what it is like to be *you*. By the same token only you know how it is to be you. Your qualitative experiences, your 'sense of self', cannot possibly register in any instrument because they are subjective. Qualia are thus beyond the scope of science and, according to Chalmers, that is the 'hard problem' of consciousness.

At the root of the hard problem that Chalmers, Nagel and others describe lies the Cartesian concept of subjectivity and its clash with the objective methods of reductionist science. Ever since Descartes separated the world of knowledge into two magisteria[11] – science and religion – science evolved and triumphed by following a specific methodology called the scientific method, which we touched upon earlier. This method uses experiment as its main tool in posing questions and discovering truths about the material world. Experiments are objective, at least in principle. As we have seen, they are repeatable and can be verified by many independent experimenters. Their outcomes are therefore independent of their observers.[12] To achieve this, experiments break down complex natural phenomena into smaller parts, on the assumption that if one is able to understand the parts one can also understand the whole, a process called reductionism. Reductionism and objectivity have been essential characteristics of science since the Renaissance and remain so today.

Imagine now an experiment in consciousness. Say you are placed inside an advanced brain scanner of the future with which the neuroscientist in charge can record, view and analyse every single neurobiological process taking place in your brain at any given moment. It is as if your whole brain can be decoded and everything that there is to know about it can be known. The neuroscientist turns on the machine: your brain scan is an objective piece of experimental evidence. You are then shown a picture of a beautiful red rose and the scanner records all that happens in your brain. And yet there is something that the scanner cannot record: your own, personal experience of the redness of the rose. For this the neuroscientist must ask you: what is it 'like' to see that rose? In reply to which question only you, the subject, can report your experience. There is no way that your personal experience can be verified objectively. You are the only one that knows 'what it's like' to see that particular rose. Thus, claim the dualists, the scientific method is insufficient to explore and explain the full phenomenon of the conscious experience. Your qualia cannot be reduced to neurobiological processes. The object (your brain) needs the subject ('you') in order to complete the picture. Therefore, 'you' are different from your brain. If you were not different you would be a philosophical zombie, a purely materialistic creature objectively reduced in the scanner, and bereft of qualia.

Philosophical zombies are another metaphor for intelligent computers. Like intelligent computers, they would conceivably pass the Turing Test but they would not 'really' be conscious. They would fake intelligence, when in actual fact they would not be self-aware of their responses, or their actions. According to modern dualists, Artificial Intelligence cannot exist in any genuine, human-like sense. What could exist are intelligent machines that behave like humans and converse like humans, but in fact they would be p-zombies.

But what if 'information' is soul-like? What if the soul, or the mind, were made of pure information, like a software program? What if the computer metaphor is not a metaphor, but a reality? What if Hawking is prescient to imagine a future in which one could download his mind into a computer, leave his physical body to rot, and live forever in a digital Eden? Could that really happen? Has it happened already in the distant past? Could we conceivably already be living inside a digital simulation created thousands of years ago, or yesterday? Is there any way to tell?

The disembodiment of information

We live in the information era. Our modern civilisation is founded on computer technology that processes data. Governments, businesses, scientists, bankers and terrorists increasingly realise that true power in the twenty-first century equals control of, and access to, data. At the same time, the computer metaphor frames our way of thinking, and how we communicate the fundamental ideas of our time. We speak of the brain as the 'hardware' and of the mind as the 'software'. This dualistic software–hardware paradigm is applied across many fields, including life itself. Cells are the 'computers' that run a 'program' called the genetic code, or genome. The 'code' is written on the DNA. Cutting-edge research in biology does not take place *in vitro* in a wet lab, but *in silico* in a computer. Bioinformatics – the accumulation, tagging, storing, manipulation and mining of digital biological data – is the present, and future, of biology research.

The computer metaphor for life is reinforced by its apparently successful application to real problems. Many disruptive new technologies in molecular biology – for instance 'DNA printing' – function on the basis of digital information. This is how they do it: DNA is a molecule formed by two sets of base pairs: adenine-thymine (A-T) and guanine-cytosine (G-C). The pairs are stacked along the double helix and their sequence defines the hereditary characteristics of the living being[13] whose DNA it is. The sequence is called the 'genome'.[14] Let's imagine these pairs as Lego pieces. In themselves, they are just pieces of boring chemistry. Put them in the right order, however, and you have the recipe for a complex, living creature.

By reshuffling the pieces of base pairs and putting them in different sequences biologists 'recombine' DNA molecules. This means that, given the basic parts (the A-T and G-C pairs), one can synthesise the DNA of any organism, if one knows the genome of that organism. Indeed, one can put together completely new organisms, in what is called 'synthetic biology'. DNA printing makes synthesising DNA easy and cheap to do by having a laser work on the bases and quickly arrive at the desired sequence.[15]

What is particularly interesting about techniques such as DNA printing is that the correct sequence is a piece of information that can be transmitted and processed just like any digital piece of information. This looks like a validation of the computer metaphor, and indeed of the Cartesian separation of body and mind. The parallels between life, computers and brains appear to be staggering. It seems that cells, bodies and brains are just dumb vessels. What makes the difference between a bunch of lifeless base pairs in a Petri dish and a living DNA molecule is, ultimately, information – or software. Not only is the metaphor very powerful in itself, but many scientific discoveries seem to confirm it.

Nevertheless, the idea that information exists beyond the material world is Platonic dualism par excellence. In our contemporary, post-industrial world information is privileged economically, socially and politically. Our global economy produces most of its value by manipulating immaterial symbols.[16] The computer metaphor extends to our personal lives and values, too. Millions of us live double lives: physical and digital. Not only are we members of digital social networks, but our personal data are hosted in numerous databases that are controlled by governments, insurance companies, utilities, banking, and so on. For many of us, our digital existence – and the rights it confers – is extremely important and vulnerable. When former CIA analyst Edward

Snowden revealed the extent of government spying by the NSA on US citizens, his revelations shook the political system of the Western world. What is the meaning of democracy in the twenty-first century when the state can keep a watchful eye on each and every one of us (i.e. our 'digital' selves)? *Who watches the watchmen*[17] in a digital world? Plato seems to have won every argument: form comes before matter. As in the *Matrix* movie, we can die in the 'real world' if we get killed in the 'digital world'. Data is life.

However, if that is true, we must also explain why the material world should exist at all. What is the purpose of molecules, atoms, planets, legs, hearts and galaxies? If we accept that information – or consciousness, or data – are more fundamental than atoms and molecules then we are surrounded by astronomical waste. Evolution furnished us with excess baggage, called bodies, which we do not really need. Apparently, all we really need is our brain, and maybe not even that. If our minds, or souls, are made of pure information (whatever that may mean) then we *are essentially* software programs. The ontology of humanness is thus reduced to information. Following this logic, the software program that codes our consciousness can be extracted from its biological substrate, downloaded on to a computer and transmitted to the end of the universe at the speed of light. We can thus become immortal and be uploaded to a higher, ethereal, digital plane of existence. Perhaps, then, the 'purpose' of a material universe is to arrive at a time when intelligent beings like us can dematerialise it, after they have first dematerialised themselves. This is a curious conclusion. There is something profoundly teleological and apocalyptic about it. In fact, it looks like a rehashed belief in the afterlife for atheists and agnostics.

At the gates of digital heaven

The Christian resurrection narrative has subtly changed over the centuries. In the past, Christians believed that the soul would return to the body on Judgement Day, and that resurrection meant the literal reunion of body and soul.[18] The dead would actually rise from their graves like zombies in the movies – only looking and behaving a lot better. St John's Revelation is quite explicit about this. Nevertheless, few Christians have faith in this scenario any more. Most believe that the soul goes directly to heaven and that the body perishes forever. Not surprisingly, many scientists with Christian beliefs do not question the computer metaphor rigorously enough: its fits very well with current ecclesiastical doctrine.

One of the most influential Catholic thinkers of the twentieth century, the French Jesuit Pierre Teilhard de Chardin (1881–1955), saw purpose in the ultimate disembodiment of the human intellect. Teilhard – who was influenced by the Neoplatonic ideas of St Augustine – believed the universe was evolving to become ever more complex and more conscious, towards a point in time he called the 'Omega Point'. In 1922, he introduced the term 'noosphere' to denote an ever-expanding sphere of human thought. Many regard the noosphere as a prophetic foretelling of the Internet. Teilhard's ideas also inform the concept of 'AI Singularity'– the quasi-religious, teleological belief that Artificial Intelligence will overrun human intelligence by mid-twenty-first century.[19] The main proponent of AI Singularity is the futurist and inventor Ray Kurzweil. He claims that, by 2045, AI will have progressed so rapidly that it will outstrip humans' ability to comprehend it. Once the Singularity has been reached, intelligence will radiate outward from the planet until it saturates the universe. The AI Singularity futuristic narrative seems like a retelling of Teilhard's Omega Point – or Judgement Day if you prefer – when the sum of intelligence in the universe accelerates exponentially thanks to self-improving Artificial Intelligence. Thereafter AI absorbs all sentience into its merciful wholeness. The verdict is out whether this would be heaven or hell.

Teilhard has also been very influential on the authors of the Anthropic Principle.[20] The Anthropic Principle tries to make sense of why the universe is so finely tuned for life to emerge and evolve. This fact is profoundly evident from the so-called 'physical constants', pure numbers that govern the natural laws. An example of a physical constant is the 'Plank constant', the number $6.62606957 \times 10^{-34}$ m^2 kg/s that shows up in just about everything concerning quantum physics. It relates the energy of particles to the frequency of their oscillation. If the Plank constant were slightly different from 6.62606957, there would not be stable nuclei in atoms, which means that atoms would decay faster than they do, and therefore there would not be complex chemical molecules. No chemistry means no life, no humans, and no AI. So how come the Plank constant is *exactly right* for life? Why doesn't it have any other value but this 'right' one?

The Anthropic Principle claims, somewhat tautologically, that the universe is finely tuned because we are here to observe it. There simply could not have been any other way. Only a universe capable of eventually supporting life could produce intelligent observers, who would then observe how finely tuned that universe is. If there have been, or are, other universes different from ours we might as well regard them as non-existent: there is no one there to observe them. This innocent sounding tautology becomes very controversial in its 'stronger' version, one that Teilhard would recognise and rejoice in: that the universe is *compelled* to allow conscious life to emerge eventually. This is where the Strong Anthropic Principle meets the AI Singularity: Kurzweil, Barrow and Tippler believe that there *must* be a purpose for intelligence in the universe. That intelligence cannot be a mere evolutionary accident that took place on a small blue planet on the outer ridges of an insignificant galaxy amongst the hundred billion galaxies that make up the observable universe. That intelligence is the pre-ordained seed of something bigger. But what could this be?

In their book *The Anthropic Cosmological Principle*, Barrow and Tippler imagine a far distant future when the universe is slowly dying. This is happening because of a law in thermodynamics called entropy. This law states that heat flows from the hotter body to the cooler one – and you can test the law yourself anytime by wrapping your hands around a warm cup of tea. Keep your hands there for a while, and the temperature between your hands and the cup will ultimately equalise. There will be no more heat transfer. Your hands and the cup will be at the same temperature. The same thing happens across the universe. Heat is transferred from hot bodies such as stars to the cold expanse of space. As we know, the universe is expanding all the time, which means that space is getting ever bigger. In the far future, and because of entropy, the energy in the universe will become dissipated evenly across the vast distances created by its expansion. Our universe is destined to die slowly from entropy. But long before it dies all biological life will have perished. And that's because biological creatures need energy to survive, and to get it they must somehow extract it from places where it is dense. Such places will become increasingly scarce in a future universe in which energy levels will be the same nearly everywhere: very close to zero. In that distant future, claim Barrow and Tippler, intelligence could only be artificial – in the guise of conscious self-replicating machines that will have colonised the universe over innumerable millennia. The ultimate purpose of this intelligent, mechanical life would be to prolong the universe's existence to its maximum extent by balancing its last remnant energy. In other words, intelligence has a purpose: it is what the universe created in order to preserve itself in old age.

If we are to take this notion seriously then we must also consider the possibility that these artificially intelligent beings of the future might already exist. The future could be *now*. Indeed, these hyper-intelligent watchmen of the universe may already have constructed a digital simulation of the universe in its youth, similar to something in a science experiment, from which to learn things as they scurry around seeking the last remnants of energy. Perhaps this digital simulation is where we now live. Perhaps we humans are actually digital programs run by hyper-intelligent AI living in a material world concocted from whatever raw materials are left in a dying universe. It is a disturbing thought that becomes even more disturbing when one realises how hard it is to tell whether it is true or not. So let's examine it a little more closely.

The idea of our living in a computer simulation assumes first and foremost that the body–mind duality is true. The philosopher Nick Bostrom,[21] one of the proponents of a simulated universe, concedes that the basic assumption underpinning the concept of a simulated universe is the so-called 'substrate-independence'. This is the idea that mental states supervene over physical substrates. Therefore, it is not an essential property of consciousness that it is implemented in carbon-based biological neural networks inside a cranium. As Bostrom says, 'Silicon-based processors inside a computer could in principle do the trick as well.' Could Bostrom be right? Or Kurzweil – or Descartes for that matter? Could we be all mind: pure information patterns that can be uploaded, transmitted and processed just like a digital document or a worksheet in Excel?[22] One way to examine the veracity of these arguments is through the perspective of Artificial Intelligence. It may appear that body–mind duality supports the evolution of AI. In fact, if you take the position of the Strong Anthropic Principle or the AI Singularity, you may think that dualism *compels* the evolution of AI. Alas, as we will see, this is far from true.

Dualistic dead ends

Body–mind dualism poses insurmountable problems for Artificial Intelligence. If we accept that the mind is independent of the brain we come to two complementary conclusions with regards to Artificial Intelligence that lead us to dead ends. Let's examine them in turn to understand why this is so.

The first conclusion, the one supported by the disembodiment of information, tells us that Artificial Intelligence will not be different in any way from the carbon-based, cranium-locked intelligence of the human variety. In fact, it tells us that AI will be indistinguishable from 'natural' intelligence in the near future and that the two will fuse. If minds are made of information and software is made of information, then minds and software are the same. If you mix up the bits, the result will be an infinite variety of human-machine cyborgs made of pure information.

The second conclusion, supported by the explanation gap and the hard problem of consciousness, tells us that it is impossible to know if a computer is truly intelligent: AI could be like p-zombies. AI could behave as if it were conscious, but we could never be certain that it had subjective experiences. An intelligent machine would pass the Turing Test, but it could not be 'truly intelligent' because it would not have subjective experiences. Nevertheless, we could never be absolutely sure that it did not have subjective experiences: we would simply not know. Since a '*noometer*', an instrument that measures subjective experiences, is theoretically impossible to create according to the adherents of the explanation gap, we can never know anything about the internal world of an intelligent machine, in much the same way that it is impossible to objectively verify anything about the internal, subjective world of any fellow human being. The hard problem – if true – tells us that we must content ourselves with knowing nothing objective about the subjective experiences of others, whether they are made of carbon molecules or silicon chips.

Combining these conclusions, we arrive at the premise that conscious AI is *hypothetically possible* but *experimentally unverifiable (or falsifiable)*. But this statement is trivial. It basically says nothing at all. The same can be said about God, angels or fairies. Hypotheses that cannot be verified (or falsified) by experiment do not belong to science but to religion. Similarly, the principle of body–mind dualism requires that we trust it on the basis of faith. It does not inform us of anything from which we could gain any useful new knowledge. If we accept that the body and the mind are two different things, we are left in the dark about both. We find ourselves back in the Upper Palaeolithic: everything is subjective.

An even more disturbing correlate of body–mind dualism has to do with digital downloads of consciousness. If the mind is independent of the body and privileged enough to exist without it, then digital downloads of consciousness are possible. However, a characteristic of digital information is that it can be copied *exactly*. One can thus make infinite, exact digital copies of oneself. But which one of these copies would be the 'true' self? Say, for example, that you are a purely digital consciousness living in a simulated universe and that you are told that a mistake was made – too many of 'you' were copied by some error – so now *you* must be deleted. Would you accept this request as morally correct? Or even logical? In a digital universe of simulated beings there is no archetypical self, no 'first mould'. *We are all copies.*

Dualism thus takes us to a paradoxical conclusion that negates the very principle of *cogito ergo sum*. Dualism ultimately undermines its own position. Instead of demonstrating the uniqueness of experience, it does the exact opposite. If the mind is made of information, it can be digitised and copied infinite times. Therefore, if we accept body–mind dualism as true, subjectivity can also be copied infinite times; it thus becomes so trivial that it might as well not exist at all. For, by infinitely copying the self, we will arrive at the sum total of all possible subjective experiences. And that would automatically translate into an objective measurement.

Perhaps, then, the mind is not really separate from the body. Perhaps the mind and body are made of the same, material substance as anything else. So let us examine this alternative, materialistic, hypothesis[23] and see where it takes us.

9

LA RÉSISTANCE

Imagine a planet where an intelligent species has evolved with a central nervous system similar to ours. Like ours, their brains have the ability to develop general-purpose language and a theory of mind. They are social creatures; they dream, love, play music and produce wonderful art. From hunter-gatherers, they gradually evolve complex agricultural societies, and invent legal systems and religion. They colonise the continents of their planet and become the dominant species. Here's the question: will that species develop an advanced technological civilisation, as we humans have done on Earth? Will they discover the laws of gravity, electromagnetic radiation, how to smash the atom, build computers and send probes into space? Is technological development the deterministic result of brains endowed with a high degree of intelligence?

Although this may seem like an impossible question to answer, there is some indirect evidence that suggests having a clever brain is probably not enough to invent computers, spaceships and antibiotics. Consider our galaxy with its one hundred billion stars. In 2013, data from the Kepler space mission[1] showed that there might be as many as eleven billion Earth-sized planets orbiting the habitable zones of sun-like stars. Assuming life is not a uniquely earthbound phenomenon, it is reasonable to expect that on many, if not all, of these planets, life must have at least appeared and possibly evolved. In 1961, the astronomer Frank Drake proposed an equation that quantified the number of possible technological civilisations in the Milky Way. His famous 'Drake equation' is the theoretical basis behind SETI, the search for extra-terrestrial intelligence. Drake's equation takes into account several factors[2] in order to arrive at widely varying estimates. Some estimates suggest that there ought to be around thirty-four million advanced civilisations in our galaxy, although this figure has been challenged.[3]

Whatever assumptions one may choose to plug into Drake's equation, the fact remains that to this day we have not received a single transmission from another planet. By definition, an advanced technological civilisation is one that has mastered and used telecommunications. Some of these potentially habitable planets lie less than a few dozen light years away from Earth. If there were advanced technological civilisations out there we ought to have been watching their television shows by now! Radio waves ought to be teeming with intergalactic chat. And yet nothing but utter silence comes from outer space. This fact is often referred to as 'Fermi's Paradox'.[4] If life is ubiquitous and evolution universal, then the sheer number of stars and planets in our galaxy ought to have produced several advanced civilisations, many of which must have been millions of years more advanced than ours. But, if so, where is everybody?

Several explanations have been offered as to why we have heard nothing so far. The astronomer Carl Sagan believed, rather pessimistically, that technological civilisations self-destruct soon after they discover nuclear weapons. Until his dying day he fervently opposed nuclear proliferation. Others have suggested that, ultimately, few planets can support complex life forms; and that the habitable planets of the Milky Way are home to mindless bacteria at best. I would like to add another explanation, one that has little to do with astronomical statistics and everything to do with philosophy.

Consider Earth: the reason why we have an advanced technological civilisation is because we have science. But science is not an idea that comes naturally. In fact, it is a rather unnatural idea. As we saw in Part I, science goes against our cognitive architecture, which instinctively assumes invisible intelligent agencies as the cause of all natural phenomena. To reason scientifically means that you must overcome your own mind. Just think of the billions – in our highly technological twenty-first century – who still believe in astrology, magical cures and supernatural beings. Arguably, only a minority of humans can accept a purely scientific, or materialistic, explanation of the world. Even many practising scientists declare themselves to be agnostic, or profess their faith in an almighty God as the primal cause of the universe. This is not surprising. Evolution has made us dualists. Our whole experience as human beings compels us to think in non-scientific ways. Some secretly, others not, but we all pray for something sometime.

This unnatural way of thinking called science did not appear everywhere on Earth, but only in a very small part of it called Europe. Although other advanced civilisations, such as the Chinese or the Indian, made important discoveries about the natural world and impressive technological advances, it was in Europe that science was first systematised as a discipline of enquiry antithetical to assuming supernatural causes for natural phenomena.[5] Science is therefore the unexpected result of many haphazard historical coincidences that took place at a particular place and time. Historicity and cognitive unnaturalness suggest that it would have been extremely unlikely for an intelligent creature with a similar brain to ours to discover – and widely adopt – science. At best, these aliens would evolve highly sophisticated pre-industrial societies. They may have developed astronomy and mathematics. Perhaps they managed to invent some basic machines. But their dualistic way of thinking could only take them so far. They would probably have someone such as Plato, whom they would consider to be their greatest philosopher. And a pantheon of great priests, messiahs and mystics. But they would not have had an Aristotle; and even if they did he would have been quickly dismissed and ultimately forgotten as an oddity.

Plato's best student

If Earth proves to be the only technologically advanced planet in the Milky Way, then this is probably because in 384 BC a boy was born in Macedonia, northern Greece, who was destined to become the first scientist in the galaxy. At eighteen years old, Aristotle joined Plato's Academy and remained there until he was thirty-seven. After Plato's death he left Athens and returned to Macedonia to become the tutor of the young Prince Alexander. They say that, as Aristotle left Athens rather hastily, he famously remarked that he was not going to allow Athenians to make 'the same mistake twice'. What he meant was that he would not let them prosecute him and condemn him to death, as they had Socrates. But why would the Athenians have wanted to assassinate Aristotle?

The most probable explanation is his intimate connection to the royal house of Macedon. Aristotle lived at a pivotal point in history. Athens was in decline and threatened by the Panhellenic rhetoric of King Philip II of Macedon. Athenians regarded Philip as an aggressive tyrant who was poised to deprive them of their freedom and democracy. They were right and soon they, as well as the rest of the free Greek cities, would be united by force under the sceptre of Aristotle's employer. Philip's son Alexander the Great would subsequently conquer the Persian Empire and extend the boundaries of his own to the Indus River. Greek political life would become a simulacrum of its past. But its arts, language and ideas would be exported to the Near East, Egypt and as far as India, in great abundance. From the cross-fertilisation that ensued, new hybrid civilisations would emerge during the so-called Hellenistic era. Aristotle's philosophy would be very influential during this new period, which lasted until the conquest of the Hellenistic kingdoms by Rome. Competing head-to-head with Platonism, Aristotelians would manage to shape the first scientific revolution of our world and encourage breakthrough discoveries in medicine, astronomy and mathematics, as well as the engineering marvels of automata, astrolabes and the first steam machines. The seeds of the West's dichotomy between form and matter, between the opposing worldviews of Plato and Aristotle, were sown at that time.

So let me examine the core differences between Aristotle's thinking and that of Plato. They both agreed on the Socratic theory of universals, or forms. They accepted that there was a hidden order in the universe. But Aristotle differed from Plato in his belief that knowledge about universals could be obtained through perception rather than pure reflection. He posited that one had to study natural phenomena in order to discover that hidden order, those Socratic universals. Simply pondering them and using logic was not enough. Aristotle was thus the father of empiricism.

But what of the *nature* of the universals? Where did they exist, and what were they made of? Plato believed that universals were uninstantiated, i.e. that they existed in an immaterial, ideal form even if they did not exist in a physical form. For example, according to Plato, natural numbers[6] exist independently of physical things being numbered or ordered. To a Platonist, mathematicians do not 'invent' mathematics; they 'discover' them, much as Columbus discovered America. Mathematics is always 'there'.

Aristotle believed the opposite: the only universals are the ones instantiated in existing, physical things. So every time we order objects we instantiate the natural numbers. For Aristotle, natural numbers do not exist unless they are predicated on physical objects. He thus gave precedence to matter over form. Aristotle's call to his students was to get out of the classroom and begin to explore the material world. He set the first example by writing treatises on just about everything – including physics, biology, zoology, ethics, logic, music, government, rhetoric, aesthetics, poetry, linguistics, psychology, geology, optics and metaphysics. It is not an overstatement to say that Aristotle, in his time, was *the man who knew everything*. For he was the first human to start compiling the body of knowledge we nowadays call science.

He was, of course, wrong in most of the things he wrote about, and for this he is often criticised, even today. But this is the thing about science: it always changes, it is constantly revised, it is never absolute. The Aristotelian body of scientific knowledge was ultimately rejected because it did not stand the test of further and deeper observation of nature. However, this is how science progresses, by being sceptical and constantly revising its ideas in the light of new discoveries and knowledge. For many people this is very hard to accept. The lack of certainty wreaks havoc with our cognitive systems. We are creatures that long for the psychological security of the absolute and the unchanging. Plato is more charming than Aristotle because Plato talks about the immutable and the eternal. On the contrary, scientists say one thing today and another thing tomorrow; and this can be very frustrating to hairless apes like us. It is no wonder, then, that science has some very serious, and unresolved, issues when it comes to communicating its ideas effectively to a general audience (and that includes politicians across the board).

But what did Aristotle think of the 'mind'? In his book *De Anima*,[7] Aristotle explores this question and suggests three kinds of souls: the vegetative, the sensitive and the rational. However, his concept of soul is very different from the Platonic concept most of us share today. The Aristotelian soul is not an eternal immaterial entity. It is the 'form' of a living being. For Aristotle, living beings are composites of form and matter. Form gives living beings their function – for example, their ability to move, which for Aristotle was central to the phenomenon of life. Growth and chemical transformations were considered types of movement, like the Hippocratic concept of humours flowing inside the body and conferring life. The rational soul of humans was capable of all the powers that the other two souls had – the vegetative's power to grow and nourish oneself, the sensitive's power to experience sensations and move – plus the power to receive forms of other things and compare them.

Thus the Aristotelian rational soul is virtually the same as consciousness: the rational mind has the ability to become aware of the world and make sense of it. However, because for Aristotle matter takes precedence over form, consciousness can only be instantiated in the body. In modern terms, this means that consciousness is a purely biological phenomenon. Aristotle was no dualist. At death, so he believed, form is destroyed together with the body. Therefore there is no consciousness left once the brain is dead. Our 'soul', or 'mind', *is* our functioning brain. Form and matter are one. There is no afterlife, digital or otherwise.

Aristotle goes viral (and mutates)

Aristotle would not have been so influential if it were not for Ptolemy, a close friend of Alexander the Great, who later became king of Egypt and began the long dynasty of Greek kings and queens that ended with Cleopatra 275 years later. Ptolemy had studied under Aristotle as well,[8] and when he took up his throne in the newly built city of Alexandria he set off to realise his teacher's dream of universal knowledge. His most eminent work was the foundation of the Library of Alexandria, one of the seven wonders of the ancient world – the Wikipedia and Google Search of antiquity rolled into one.

Under the Ptolemies, Alexandria became an international centre of learning that specialised in astronomy, medicine, mathematics and engineering. Following the Roman and Byzantine conquests of Egypt, the arriving Muslim armies of the Arabs took up the Aristotelian science for which Alexandria was famous, and expanded it further. Averroes[9] and Avicenna[10] wrote and commented extensively on Aristotle's work, and Muslim theologians considered him the 'First Teacher'. By way of Spain and the Moors, Aristotle arrived in Western Europe towards the beginning of the twelfth century. There, as we saw, largely thanks to St Thomas Aquinas, Aristotle displaced Plato as the principal source of philosophical discourse for the Catholic Church. And so it was that Aristotle gained firm roots in Western European thought. The inquisitive minds of the Renaissance and of the Enlightenment looked up to him as they began to study nature, cataloguing it meticulously, and conducting experiments with their new observational instruments. And that was how modern science was born.

Then Descartes came along and – as we saw – brought Plato in through the back door. As a reaction to continental rationalism,[11] the English philosopher and physician John Locke (1632–1704) kick-started British empiricism by reasserting the Aristotelian dictum that knowledge is based on experience as evidenced by the senses. Locke suggested that in order to gain knowledge about the material world one had to build hypotheses that were testable by observation and experiment – the approach that we refer to today as the scientific method. Soon it became apparent to religious circles that Locke's empiricism was a one-way street to atheism. To save religion, the Anglican bishop George Berkeley (1685–1753) confounded empiricism by pointing out that since our senses are the prime source of knowledge then things exist only as a result of their being perceived. God was thus reinstated since He was the one who created the human mind.

Berkeley's ideas were later secularised and became very influential in modern philosophy, as the basis for doubting the existence of objective reality.[12] Scottish philosopher David Hume (1711–1776) also subverted scientific empiricism by questioning logical induction, i.e. the way in which scientists validate their hypotheses on the basis of experimental results. Hume noted that it was ultimately impossible to link cause and effect. He therefore claimed that we could never be certain that the future will resemble the past. You may do an experiment a million times and get the same result, but that does not mean that you will inevitably get the same result if you repeat the experiment one more time.[13] Hume and

Berkeley ultimately arrived at similar conclusions: that all physical objects are constructions of our experiences. Which meant that only mental objects existed. This idea infuriated many people at the time, including the famous English writer Samuel Johnson (1709–1794). When he was told of Berkeley's proposition that the material world was non-existent he struck his foot with mighty force against a large stone and shouted, 'I refute it thus!'[14]

Nevertheless, questions concerning the ability of the senses to derive knowledge from the physical world continued to worry philosophers. As the twentieth century dawned, it became apparent that performing experiments was problematic because this assumed an idea of 'normality'. As we saw earlier, the validity of experiments as means of testing scientific hypotheses rests in their repeatability. A fundamental condition for repeatability is that all experiments should be carried out under the same conditions – the so-called 'normal conditions' – and by 'normal' observers. But who can decide who is 'normal'? Perhaps a doctor could be called in to make sure that the observers of the experimental results are 'normal'? But then one would need another doctor to make sure that the first doctor is also 'normal', and so on . . . The problem lies not so much in the absurdity of finding an infinite sequence of doctors to check on each other, but in the realisation that the description of the world is always contingent on the species which does the observations, experiments, and so on – in this case us humans. And that's because our senses, as well as our sense, is a product of the particular line of evolution that begot our species. 'Normality' is therefore neither unbiased nor universal and, therefore, cannot be trusted. The serious problem with a species-specific sensing apparatus (our brain) was compounded further with the quandary of describing scientific observations using our very imprecise, general-purpose language. Science seemed to be deep in trouble.

A way out of this mess was explored in the Roaring Twenties

by the neopositivists of the Vienna Circle. They attempted a synthesis of British empiricism and mathematical logic. Their big idea was to get rid of the nuances and uncertainties of natural language (and therefore of the subjectivity of the human observer) and to replace this with a perfectly logical and unambiguous mathematical language. For them, metaphors were anathema. Risking anachronism, we can think of the neopositivists as wanting to replace humans (or any other intelligent alien species) with computers, and human language with computer language.

The neopositivists believed that it was feasible to bypass the problem of defining normality when reporting experimental results. And they had two excellent reasons to back their belief: the German mathematician Gottlob Frege (1848–1925) had shown that all mathematical truths are logical, while the Austrian philosopher-star Ludwig Wittgenstein demonstrated that all logical truths are linguistic tautologies, i.e. they are true in every possible interpretation (and therefore immune to subjectivity). Putting the two together meant that only logic and mathematics guaranteed an exact description of empirical data about the physical world. Everything else was chit chat. In fact, Wittgenstein and others arrived at the conclusion[15] that sentences that were not purely logical or unverifiable were indeed devoid of meaning. Metaphysics, ethics, aesthetics and the like were pseudo problems, and therefore unworthy of study. For the neopositivists it was better to stop talking altogether about that which was not purely logical and verifiable. As Wittgenstein famously quipped, 'What we cannot speak about we must pass over in silence.'

And yet any problems concerning the use of empiricism as the best way to explore reality were not settled once and for all by the brilliant neopositivists. As we shall see, logic has its limits – a fact proved by Kurt Gödel, a member of the Vienna Circle. Moreover, twentieth-century science discovered physical objects, such as genes and electrons, which are not directly detectable to the human senses. The infinitesimal microcosm of nuclear physics and molecular biology requires that we trust our measuring instruments and create abstractions in place of actually sensed physical objects. No one has ever seen an electron. We do not really know what it looks like. The deeper we go into physical reality the more it acquires an intangible aura that leads many thinkers towards the denial of objectivity and the return of Platonic anti-realism.

Thankfully, as philosophy kept itself busy debating the limits of empirical science, science moved on. Science and the scientific method did not simply bring new knowledge about the world. They ushered in a new age of engineering invention and, ultimately, the Industrial Revolution. No matter how interesting or valid philosophical scepticism about science is, human societies have benefited from the results of scientific and engineering research. The study of the human body begot modern medicine. The study of microorganisms led to new drugs that saved millions of lives. Ballistics determined the victors on the battlefield. Steam engines, ships and railways laid the foundations of empires, and changed the route of history. Aeroplanes took to the air. Men walked on the face of the Moon. Most people on Earth today live longer and enjoy a better and healthier life than every other human generation that has lived on this planet before. Empirical science – somehow and despite its many philosophical flaws – seems to work very well indeed. In fact, one may argue that the sole reason science is still a dominant part of our culture lies in its practical applications, the engineering of life-saving or life-enhancing solutions. Our survival and wellbeing benefit from science, and that is why we respect this cognitively unnatural thing. But could all these tangible achievements of science and engineering be mere constructs of our minds? What about that arch-creator of constructs, dreams, philosophies and science – the mind? Can empirical science tell us anything about that? Can science explain consciousness?

Consciousness explained

Daniel Dennett is a philosopher with a mission to infuse some empirical sense into the problem of consciousness. I met him in Tucson, Arizona, in April 2006, when I interviewed him for a popular science magazine I used to edit. A stout and affable academic who enjoys talking with non-experts, and with a distinctive white beard, radiant blue eyes and a keen sense of irony, Dennett is perhaps the closest to a modern version of an ancient Greek philosopher one can get. As you may imagine, he thinks very lowly of Plato and very highly of the Presocratic materialists and of Aristotle. His views on the mind problem have upset many of his colleagues in the dualist camp – particularly his dismissive take on regarding subjective experiences (qualia) as the 'hard problem' – which he elaborated upon in his 1991 landmark book, *Consciousness Explained*.[16]

Dennett refuses to accept that the mind is mysterious and somehow beyond the scope of science. He often uses the analogy of magicians who trick our minds into believing impossible things. The lady on stage is never sawn in half – and yet it always looks that way. Likewise, he claims, our brain convinces us that we have 'consciousness'. Qualia – or subjective experiences – are therefore magical illusions created by material processes in the brain. There is no 'hard problem': it is only a matter of time before empirical science will explain subjective experiences, as it has done with many other phenomena previously considered 'mysterious'.

At the core of Dennett's empiricism lies the notion that mental phenomena are *identical* to neural processes taking place in the brain – an idea often termed 'identity theory'. This notion is similar to saying that lightning is *identical* to electrical discharge, or water is *identical* to two molecules of hydrogen bound with one molecule of oxygen.[17] The language we use to describe a certain phenomenon should not confuse us. We talk about 'consciousness' as if it were something unitary, when in fact it is the result of multiple interactions taking place at the level of 'unconscious' molecules. Identity theory is thus an idea driven by scientific development and the reductionist credo that, ultimately, all natural phenomena can be reduced to interacting elementary parts. In the case of consciousness these elementary parts would be the neurons and their connections.

Although Dennett is not a neuroscientist, he has attempted to put together a comprehensive, philosophical theory of consciousness, which could provide a model for, or an interpretation of, scientific findings. In his public lectures he often uses set-ups from cognitive psychology experiments. These experiments aim to identify and meaningfully understand discrete mental states in the brain. One of these mental states is 'awareness' – the moment we actually become aware of something. For instance, there is an experiment during which a subject is shown two slightly different photographs in quick and repetitive succession, and is asked to spot the differences in them. Dennett uses such experiments in order to demonstrate that it takes us several seconds to become aware of the differences between the two similar photographs. And yet the information from the photographs has arrived at our brain almost instantaneously. What happened between the time of information arriving and us becoming aware – or conscious – of the difference between the two photographs?

Dennett, as well as many cognitive psychologists, argues that the human brain processes information without our being conscious of that fact. For Dennett, the brain creates several versions of what is going on out there, like an author writing several drafts of the same story.[18] Consciousness is when one of those drafts becomes the dominant one. In other words, the mental state we call 'awareness' is identical to the end of a process during which our brain has edited out all other drafts of what is happening, leaving but one.

The driver of the editing process in our brain is intention, which is as good an explanation as one can get as to why we have consciousness in the first place. It is because intentions are fundamentally linked to our survival. To understand why this is so, let's think of Dennett's experiment in somewhat different circumstances. Instead of two slightly different photographs flipping in the safety of a psychologist's lab, let us imagine ourselves a million years ago in the African savannah watching out for lions in the bush. The wind sways the grass and our vision constantly processes the visual information our eyes receive. As a result of evolution, humans and other animals do not look at a scene in fixed steadiness (as most birds do). Our eyes move around, locating interesting parts and building up a mental, three-dimensional map corresponding to the scene.[19] This 'saccadic', as it is called, movement of the eyes is unconscious and it happens because the central part of our retina that provides the high-resolution portion of our vision is very small. If you put your arm out in front of you and focus on your thumbnail, this is how small your eyes' resolution is at any time. Saccades are the fastest movements produced by the human body; the peak angular speed during a saccade can reach 900^{0} per second.[20] Back in the savannah, our primate eyes would transmit visual information to our brain in a constant 'spot the difference' game, in case a lion emerges from the grass. If a lion does appear, then those multiple drafts of what is going on would quickly have to be collated into a single moment of awareness ('gosh, a lion!') that would prompt an urgent flight-or-fight decision ('let's climb up that tree!').

Dennett recognises a de facto evolutionary aspect to consciousness. After all, if one accepts an empirical approach to the mind problem, it could not be any other way. If consciousness is a biological phenomenon then it must have evolved over time. Creatures with simpler nervous systems than primates should exhibit longer moments of awareness; and animals without a central nervous system no awareness at all, their flight or fight reactions being purely automatic. This is a testable hypothesis that adds intellectual weight to the empirical approach. It also argues that 'meaning' is a value system that has evolved in a Darwinian manner. Our meaningful, subjective experiences are therefore identical to the evolved self-organisation of mental states in our brain.

Dennett's reductionist theories seem to support the computer metaphor of the brain. The grey matter in our skulls can be reduced to an information-processing machine. But what exactly *is* the information that our brain processes? What is information made of? We saw in the previous chapter how eminent scientists such as Stephen Hawking believe that, one day, consciousness could be uploaded in a computer. Could it be that the brain is a material machine processing immaterial information bits? Could the self be something different from the brain – for instance, a complex pattern of bits? Does materialistic empiricism lead us back to the arms, or fangs, of non-materialistic dualism?

So what is information, *really*?

We live in an era of disembodied information. Movies, books, Silicon Valley tycoons, futurists, geneticists and computer engineers posit that we have come across a new force in the universe called Information, and that this force will guide not only our destiny but answer all the big questions, including how the universe came to be, what life is and how we have consciousness. Not a few claim that it is theoretically possible to ditch our biological bodies and take immortal flight, existing as pure bits in cyberspace. Our contemporary cultural milieu is defined by a Platonic precedence of information (read 'form') over matter. As the postmodern literary critic Katherine Hayles notes,[21] we regard information 'as an entity distinct from the substrates carrying it; a bodiless fluid that could flow between different substrates without loss of meaning or form'.

This suggests that we have ceased to be unitary persons. Since the last two decades of the twentieth century we have become 'posthumans' living two parallel existences: one embedded in the material world and one in cyberspace. These two existences are rarely in sync, and are often governed by different sets of laws. For example, laws in every country of our planet prohibit the buying and selling of human beings, a practice known as slavery. However, when it comes to our informational selves – our data, our genes, our Facebook timelines – no such strict restrictions apply universally. Our digital selves are bought and sold every millisecond on Google. 'Privacy' has been redefined in terms of access to our personal data rather than our right to seclude our material bodies from society if we so please. Our matter (our physical body) has become culturally, politically, legally and economically separated from our form (our personal digital data). How did we come to this?

The notion that information is distinct from molecules and atoms goes back a long way, to the beginning of the information revolution. In his 1948 landmark book *Cybernetics*,[22] Norbert Wiener writes: 'information is information, not matter or energy. No materialism which doesn't admit that can survive in the present day.' In a strange twist of history, it appears that after the Second World War computer scientists and engineers dismissed Aristotelian empiricism and became Neoplatonists. Instead of adhering to the principle that universals are always predicated by physical objects they adopted the opposite, Platonic, idea. In Part III of this book, we will explore in more detail the technological reasons behind this conceptual shift. But before we do so, let us first examine the concept of information.

The father of information theory is the American mathematician, electronic engineer and cryptographer Claude Shannon (1916–2001). He worked as a cryptanalyst in the Second World War, and in early 1943 he met Alan Turing, who had been posted to Washington to work with the Americans on breaking the German naval codes. Like his English counterpart, Shannon is one of the great heroes of computer science, a man whose work has shaped the world we live in. He was a rather playful person, too. In 1950, he created a magnetic mouse that moved in a maze or labyrinth of twenty-five squares. Fittingly, he called it Theseus. 'Shannon's mouse' is considered to be the first artificial device with the ability to learn, and one of the first experiments in Artificial Intelligence. He likewise produced a number of other humorous inventions including rocket-powered flying discs, a motorised pogo stick and a flame-throwing trumpet. But the one I like best is something he called the 'Ultimate Machine'.[23] This was a box with a single switch on its side. When you flipped the switch, the lid of the box opened and a mechanical hand reached out and turned the switch off, before retracting into the box. I sometimes think of this useless automaton as symbolising the perennial critique of Artificial Intelligence research: no matter how clever the machine you build, no one will accept that it is truly intelligent.

When Shannon was not building humorous contraptions he was trying to solve the problem of transmitting messages over a noisy channel. We are all familiar with the concept of noise in everyday life; it occurs when a sound prevents us from listening to what we want to listen to. Noise could be the sound of an ambulance tearing past with its sirens blasting away while we're trying to talk to someone on our mobile. In generalising terms, noise is what happens when we cannot make sense of something, and therefore denotes the absence of meaning. Noise is thus a signal, or something we sense, that is bereft of information. Now imagine listening to meaningless white noise over the radio, a continuous and colourless buzz, when all of a sudden something changes in the pitch of that noise. This unexpected, or random, singular change can be said to represent 'information'. It may not have added any more 'meaning' but it *might* represent the first step in the right direction: the 'probability' that something meaningful may follow has increased. Imagine now that this slight change in pitch gradually develops into a more varied range of modulations until you begin to make out a human voice speaking in the background. The amount of information has increased rather considerably now ('someone is speaking'), although you still cannot understand what they are saying. But, beyond a certain threshold of actual signal over noise you *will* begin to understand what that person is talking about, providing they speak the same language as you. Shannon showed how one could improve the efficiency of information transmitted over noisy channels by means of *coding* the information.

Coding means representing information. For example, written language represents spoken language. One can also represent spoken language by using a code other than the usual established letters, or by using the letters in another order and so on, and thus 'encrypt' messages so that only those who possess the 'key' to the code can understand them. One way to code information is by using binary arithmetic. This is what digital information machines, such as computers, do. They convert signals that exist in the physical world into binary representations of '0s' and '1s'.[24] In binary code '0' denotes the absence of a signal and '1' the presence of a signal. Every time you use your smartphone to take a picture, light captured by your phone's camera is converted into binary digits and stored in the memory. Digital information is a long, long sequence of zeros and ones.

Shannon's breakthrough idea in his seminal paper 'A Mathematical Theory of Communication'[25] was to borrow the probabilistic mathematics of thermodynamics and apply them to the new field of telecommunications. Thermodynamics describes how molecules move as they heat up or cool down. The greater the heat, the more energetic the molecules become. A key concept in thermodynamics is how ordered the system of molecules is, or how evenly they are spread around at a given temperature. This is called *entropy* – a concept we encountered in Part II, chapter 2. When we add energy to a system – say we heat up a container filled with gas – gas molecules start speeding around unpredictably.

The order to the system increases, and entropy decreases. Using thermodynamics, we can predict not only how gas molecules will move but the fate of the universe as well. As the universe expands it cools down and thus its entropy increases; the cosmos gradually becomes ever more disorderly. Ultimately, it will reach a point of near uniformity where whatever matter or energy is left will be almost evenly spread across the vast expanse of time-space. And that will be the end of it. Our universe will arrive at a state of *thermal equilibrium*, meaning there will be no heat or energy transfer from one place to another. There will be no biological life left then, because life needs the transfer of energy in order to exist, and therefore needs a certain measure of disorder. In information systems entropy is defined as the amount of uncertainty about information. The more noisy (or more 'entropic') a message is, the less information it conveys. Shannon's information theory represents one of the greatest paradigm shifts in the history of science.

Nevertheless, in the development from thermodynamics to information theory something seemed to have gone awry. While thermodynamics described *physical* objects, information theory acquired – almost from the very beginning – an immaterial aura. Bits are not 'real' in the way that molecules are, so information was thought of as describing non-physical objects. Shannon was initially very reluctant to accept Norbert Wiener's notion of information as something beyond matter and energy. Working as an electronics engineer, he was fully aware that telecommunications are based on physics, not metaphysics. Communication signals are waves of electromagnetic radiation travelling close to the speed of light. This radiation interacts with the mechanical apparatuses of transmission and reception in very real ways: it heats them up. Digital transmission of information does exactly the same. That's why your computer becomes hot when it's on. Energy dissipates as heat in the hardware because of electrons travelling through the integrated circuits. In fact, this is the actual connection between thermodynamics and the transmission of information. Information is therefore *always* instantiated in physical objects. Whether it is our neurons perceiving the hungry lion in the bush or computer circuits processing an Excel spreadsheet, some physical substrate *always* instantiates what we can ultimately represent in bits. Zeros and ones do not exist outside the physical world. They are a representation of energy flows in physical substrates. That is why pure information is fundamentally meaningless. It acquires meaning only in context. Context affects meaning. And context needs thinking minds. Take, for example, the words 'it rains'. If I were to utter this in response to your asking me from an office in New York what the weather is like in London, then I provide you with information about something you do not know. If, however, you and I wait at a bus stop in Hackney with the sky pouring over our heads, my uttering the same phrase would probably sound moronic.[26] There would be nothing in what I had just said that you were not already aware of. Our minds contextualise and give meaning to information all the time.

An alternative way to understand the significance of context would be that any piece of information is always instantiated in physical objects: in the rain falling in Hackney, you sitting next to me or in your office in New York, etc. This is the Aristotelian and empirical view of information. And yet in both cases – and in infinite other cases that we may imagine – the *coding of information* is exactly the same. This was exactly what prompted Norbert Wiener to insist that information should be considered as something beyond matter and energy. He was frustrated about context impeding the progress of information theory. He wanted to get rid of context in order to develop a theoretical framework of information that would be applicable in *any* context. He therefore de-contextualised information as a physical, material instantiation and reduced it to the coding of information, i.e. the mathematics of information. Information was thus 'disembodied', was decoupled from physical minds giving it 'meaning', and gradually became the closest thing we have today to a Platonic universal. Several decades after Wiener and Shannon, we take it as given that information 'exists' on its own and independently of the physical objects it refers to. We have thus confused the *coding* of something with the *actual* something; the reflection of an object with the object itself. And we have thereby arrived at the dubious belief that information instantiates physical objects rather than the other way around.

Time, then, to sober up and bring together the significance of context, information theory, as well as an empirical explanation of consciousness, in order to see what an Aristotelian view of Artificial Intelligence would be like . . . .

An empiricist's primer on Artificial Intelligence

In order to see how machines can be made intelligent, we must accept the following four empirical propositions. First, that there is no soul, or spirit, or mysterious uninstantiated universals. We must reject dualism in all its versions because it leads us nowhere without explaining anything satisfactorily. Secondly, that there is only matter. That we live in a material world, and that's that. Thirdly, that intelligence, from its simplest manifestation in a squirming worm to self-awareness and consciousness in sophisticated cappuccino-sipping humans, is a purely material, indeed biological, phenomenon. Finally, that if a material object called 'brain' can be conscious then it is theoretically feasible that another material object, made of some other material stuff, can also be conscious. Based on those four propositions, empiricism tells us that 'strong AI' is possible. And that's because, for empiricists, a brain is an information-processing machine, not metaphorically but *literarily*. We have several billion cells in our body.[27] If we adopt an empirical perspective, the scientific problem of intelligence – or consciousness, natural or artificial – can be (re)defined as a simple question: *how can several billion unconscious nanorobots arrive at consciousness?*

Each one of our cells is unconscious and can be considered a 'nanorobot' in the sense that it is a mechanical automaton on a very small scale. Molecular biology and neuroscience have made great advances in understanding how cells work, and indeed how the cells in our brain (the 'neurons') work. Across medicine, a new worldwide research effort is currently underway that aims to 'translate' knowledge from biochemical research, regarding the molecular mechanisms of single cells, in order to address macroscopic health problems. For instance, understanding how RNA affects the production of proteins in the heart can help doctors invent new medicines to cure heart failure. This induction, from how single parts function to how systems of parts function, takes place in the study of neural cells, too. It is therefore only a matter of time before we manage to answer the question of how several billion unconscious automata (call them 'neurons' for biological brains) arrive at consciousness. Once we do, we will have a scientific explanation of consciousness. We will have explained the mind as we have explained metabolism, blood circulation or the birth of children. Such a scientific theory of mind, when applied in engineering intelligent artefacts, will beget conscious machines. The problem of Artificial Intelligence is therefore identical to the problem of neuroscience: understanding the transition between non-intelligent parts to an intelligent whole.

So let us, then, turn to neuroscience and see what it has discovered so far about the brain and the mind – and what lessons can we learn and apply to the creation of machines that can think.

10
PEERING INTO THE MIND

George Miller, the founding father of cognitive psychology, wrote in his 1951 book *Language and Communication*: 'Consciousness is a word worn smooth by a million tongues ... maybe we should ban the word for a decade or two until we develop more precise terms for the several uses consciousness now obscures.'[1] His words articulated the official ban on consciousness as part of any scientific investigation that had held sway since the eighteenth century. Science, flourishing under the Cartesian separation between *res cogitans* and *res extensa*, persisted in focusing on the latter, corporeal world of material things and phenomena, leaving the mental to the clergy and the philosophers. Nevertheless, and despite his aphorism, Miller's contribution to changing the established consensus was decisive. He challenged behaviourism, the prevailing theory of twentieth-century psychology, which regarded the mind as a black, uninteresting box, and focused only on observable behaviour. Adopting an Artificial Intelligence perspective, Miller tried to explain behaviour as a sequence of stimulus-response actions. In doing so, he introduced the concept of mental process, or mental function. The black box of the human mind was suddenly opened to scientific enquiry. Questions such as which mental events take place that *cause* us to remember, believe, want or feel were now asked. Combining computational theory, Shannon's informatics and linguistics, Miller created the new field of cognitive psychology. The new scientific field devised a series of experimental techniques, which would be adopted in due course by researchers of consciousness. Still, well into the early 1990s Descartes' ghost continued to haunt university psychology and neurophysiology departments, keeping them mostly separate. No serious researcher from either camp would dare to tackle the subject of consciousness.

And then, in 1995, Francis Crick, a science icon and a co-discoverer of the molecular structure of the DNA, published a book that changed everything almost overnight. Crick advocated – in very Aristotelian terms – a 'scientific search for the soul'. Consciousness, he claimed, was '*. . . entirely due to the behaviour of cells, glial cells, and the atoms, ions and molecules that make them up and influence them*'.[2] This materialistic proposition was so revolutionary in the face of the centuries-old established Platonic and dualistic concept of the mind that Crick felt obliged to entitle his book *An Astonishing Hypothesis*.

Crick's book let the genie out of the bottle. And although mind philosophers tried to defend their territory – and some still do – rigorous scientific enquiry has increasingly scaled, and captured, many of the once impregnable towers of the mystery of mind. Over the past twenty years, neurobiologists, neuroscientists and cognitive psychologists have joined forces to bring consciousness into the science lab. And what they have discovered is more astonishing than Crick could ever have imagined.

There are three principal reasons for consciousness having become an area of serious scientific research: a concise definition of consciousness, the experimental manipulation of subjective experience and the advent of powerful observational instruments.

Unbundling consciousness

In response to Miller's forbidding aphorism, quoted above, scientists have at last 'unbundled' the mixed meanings of consciousness, and articulated a better definition – one that offers itself to rigorous, empirical investigation. Nowadays, most researchers would distinguish three states of consciousness.

The first state would be the degree of wakefulness, or vigilance. This state varies as we fall asleep, or as we wake up. Once we are awake, our consciousness enters a second state called 'attention', focusing our mental resources on to specific stimuli. It has, however, been shown that we can be attentive without becoming conscious. Our brain receives massive amounts of sensory information during our waking hours and applies a filter to select what is important with respect to our goals. It then amplifies those stimuli and makes us aware of them. To visualise how this happens, imagine yourself at a party full of noise and people, trying to have a conversation with an interesting person you have just met. While you are having the conversation, and are fully focused on your interlocutor, your brain processes the information that is constantly being received by your senses. You are 'attentive' of your surroundings without being overly aware of them. If, however, someone suddenly shouts 'fire!' your attentive brain will select this crucial piece of information out of the rumbling background noise of mixed conversations, loud music, smells and whatnot, and quickly bring it to the fore of your awareness. Whenever the selected information enters our awareness it becomes reportable to others. In this 'third state' of consciousness the objective becomes subjective. Consciousness is when information becomes a story, our story.

Attention thus acts as a gateway to the third state of consciousness, which we may call 'awareness'. For example, we can consciously focus our attention on a number of sensory stimuli while looking for something specific. Think of a hunter in Palaeolithic Europe tracking deer in a dense forest, listening attentively to every cracking noise, training his eyes on hints of movement in the undergrowth.

Redefining consciousness in these three distinct states – vigilance, attention and awareness – helps neurologists distinguish locked-in syndrome in comatose patients (when the patient has conscious awareness but cannot move) from a purely vegetative state. However, it is important to note that vigilance and attention are necessary but not sufficient for awareness. For example, some stroke patients whose visual cortex is impaired become colour blind: they have vigilance and attention but lack the conscious experience of colours. Although for most people 'reality' includes colours, colour-blind patients have no access to this part of reality.

But what about self-awareness? What about the 'I' of the narrator? Research in consciousness has revealed that self-awareness is no different from becoming conscious of a colour, or a scent. Exactly the same neurobiological processes take place in the brain in every case. In fact, self-awareness turns out to be less special than we originally thought, or might have liked it to be. Indeed, there are many instances when our consciousness achieves a heightened state and the self becomes background noise, or disappears completely. Think of an orgasm, or being deeply in love, or listening to beautiful music, or gazing at a wonderful sunset. 'Mindfulness', as defined by Buddhist meditators, is a state of mind when awareness of the world occurs without the intervening 'I'. No wonder that their meditation techniques, as well as those of other religions, are also under scientific scrutiny today.[3]

The second, very significant, reason for consciousness becoming a scientific subject of investigation concerns the discovery that consciousness can be experimentally manipulated. Once attacked by science, the hard problem of consciousness turns out to be rather easy and straightforward. Subjective experiences – the qualia – *can* be turned into raw, objective data. Going back to the three states of consciousness, scientists can nowadays observe, and measure, the transition from non-conscious attention to conscious awareness. The transition event that distinguishes objective physical stimulation from subjective perception *is* consciousness. Consciousness is instantiated in the brain by physical processes. Mental processes, as studied by cognitive psychologists, correspond to neuronal processes. Therefore, the search for identifying the subjective experience can be limited to identifying the characteristics of neuronal transition events: what happens in the brain when information enters our conscious awareness and becomes reportable?

Using modern brain imaging methods, neuroscientists can observe and record what happens when the unconscious stimulus travels in the brain, including in which region of the brain it stops – and how the stimulus defines patterns of neural activity that are exclusively associated with conscious processing. They can thus establish a cause-and-effect relationship between neural processes and subjective experience. These physical processes – the transition events that *cause* conscious awareness – have been called the 'neural correlates of consciousness' ('NCC') by Francis Crick, and more recently described as the 'signatures of consciousness' by the cognitive neuroscientist Stanislas Dehaene.[4] Their discovery has become possible thanks to modern measuring instruments that can peer into the brain and effectively 'read thoughts' or, as the case often is, illusions.

Noometers[5] and vanishing gorillas

In 1990, the Japanese physicist Seiji Ogawa and his colleagues[6] invented functional magnetic resonance imaging (fMRI), a technique that enables the visualisation of brain function. We can now take colourful pictures of our thinking brains thanks to these incredible machines. But how is this possible? fMRI exploits the coupling of brain cells with blood vessels. Whenever a neuronal circuit increases its activity, the glial cells that surround those neural cells sense the surge in synaptic activity. To compensate for the heightened need for energy consumption, they open up the local arteries to let oxygen-rich blood in. Two to three seconds later, the blood flow around the activated neural circuit increases, bringing more oxygen and glucose. Ogawa's deep insight was to realise that this presented an opportunity to track blood flow using magnetism. Like a latter-day adherent of animal magnetism, he suggested placing a great magnet around a living brain and measuring distortions in the brain's magnetic field. Those distortions are caused by haemoglobin, the oxygen-carrying protein in blood cells. Whenever haemoglobin carries oxygen it acts like a small magnet; otherwise it does not. Modern magnetic resonance machines, by applying this simple principle, measure neuronal activity in every piece of brain tissue at millimetre resolution several times per second. The modern researcher of consciousness can thus peer into the brain and track how mental events arise and are propagated there. Nevertheless, there are limitations: fMRI cannot track what takes place at the level of the neuron, for instance how long firing takes place at the synapses. Measuring the time-course of neuronal firing is very important as well. After all, if consciousness is a biological phenomenon then one should begin with the biological unit that causes it: the neuron.

Thankfully, there is a technique which does exactly that, and which is called electroencephalography (EEG). Its long history dates to the late nineteenth century, when it was discovered that the brain exhibits electrical activity. The first human EEG was recorded in 1924 by the German physiologist and psychiatrist Hans Berger (1873–1941). Berger also invented the electroencephalogram, that well-known device in which many electrodes are worn around the head and which has been photographed numerous times for newspaper and magazine articles about the brain. Modern EEG machines have 256 electrodes that provide high-quality recording at millisecond resolution.

Another technique that uses the magnetic fields produced by electrical currents occurring naturally in the brain is magnetoencephalography (MEG). By measuring minuscule magnetic waves that accompany electrical discharge in cortical neurons, MEG is more precise than EEG. Using both, and in conjunction with fMRI, researchers can correlate what happens at the level of a single neuron, as well as how groups of neurons communicate and propagate information to ever wider areas of the brain during the conscious reporting of a subjective experience. These measurement methods and technologies have opened a window into the human mind that was inconceivable only two decades ago.

Given the clear definition of consciousness, clever experimental strategies were devised that isolated the perceived stimulus (i.e. the one that the brain became aware of) from the unperceived (i.e. the one that did not enter consciousness but was only attended to) by creating a minimal contrast between them.[7] Thus, one can have a pair of experimental situations that are minimally different, but of which only one is consciously perceived. EEG, MEG and fMRI can then be applied to record what has changed in the brain between these two situations. Researchers have found dozens of ways to do exactly this, inventing experimental designs that can manipulate consciousness, like a magician.

The most famous experiment to use the above strategy in order to illustrate the relationship between attention and awareness involves an invisible gorilla. Harvard psychologists Dan Simmons and Christopher Chabris won the 2004 Ig Nobel Prize[8] for making a short film that shows several people throwing a ball to each other.[9] You, the observer, are asked to count how many times the people in the film throw the ball. The experiment lasts only a few seconds.[10] You are then asked how many throws you counted. And then, whether you noticed the gorilla. Most of those who undertake the experiment do not notice a person dressed like a gorilla who walks through the people playing with the ball, stands and thumps his chest, then walks off stage!

What this experiment illustrates is that, although our brain perceives the gorilla, it does not inform us (i.e. our consciousness) about him, because we are paying conscious attention to another event (counting throws of the ball). This is no different from what happens when we drive a car. Most of our driving takes place automatically, without our being consciously aware of our driving actions. It is only when something extraordinary happens that our brain informs our conscious perception, so we can take evasive action. Given that consciousness occurs around one-third of a second after an actual event is perceived by our senses, driving and doing something else at the same time – for instance texting – can be deadly. Like the invisible gorilla, a child crossing the street while we drive and talk on our mobile might be completely invisible to us.

Other experimental designs make use of binocular rivalry: if each eye is shown a different image our perception alternates between them in sequence. Using fMRI, EEG and MEG, researchers can track the entire sequence of brain activation, from the neuron to whole groups of neurons, as the visual stimulus travels from the retina to frontal cortex, while our perception alternates between the two images. Thirty years after Crick's astonishing hypothesis, neuropsychologists have forced the human brain to reveal many of its secrets. Consciousness is not such a mystery any more.

Becoming conscious

Stanislas Dehaene is a professor of experimental cognitive psychology at the Collège de France and one of the top researchers into the neural mechanism of consciousness today. By deploying the combined arsenal of advanced technology measurement hardware and innovative experimental setups, he and his team have discovered four significant 'signatures of consciousness'[11] that take place in the human brain whenever information enters our conscious awareness.

To understand Dehaene's approach to consciousness we must first become acquainted with a fundamental tenet of modern neuroscience: that the brain is an autonomous system. What this means is that the brain can think without external stimuli. We can be isolated in a dark, soundproof room and still have thoughts and feelings.[12] This happens because spontaneous global patterns of neuronal activity originating from within transverse our brain constantly. Neurons self-activate in a partly random fashion, and then the brain system begins to self-organise. That is how we get what the American father of psychology William James referred to as the 'stream of consciousness': the flow of uninterrupted and loosely connected thoughts, primarily shaped by our inner goals and only occasionally seeking information from the senses.

Dehaene's experiments involved external visual stimuli. Vision is the favourite sense for most neuroscientists. The brain's anatomy and neural pathways of vision have been thoroughly mapped, and whether it is a vanishing gorilla, binocular rivalry, or colour-blindness, vision offers ample opportunities to test the transition from the non-conscious to the conscious. Dehaene discovered that when enough brain regions 'agree' about the importance of incoming sensory information, they synchronise into a large-scale state of global communication. This state is characterised by a burst of high-level neural activation. It is as if our brain is a large orchestra where cacophony rules most of the time, as each musician tunes their own instrument in a different way. Then, as if some musicians have managed to find a common note, more and more musicians begin to harmonise with them. The note amplifies like a surging wave that sends ripples everywhere. This self-amplifying avalanche of neural activity ultimately ignites many brain regions into a tangled state. This is the first signature of consciousness according to Dehaene: a threshold of activity is crossed and the brain activity invades many additional brain regions, leading to a sudden ignition of parietal and prefrontal circuits. The parietal lobes of our brain, at the top of our head, are concerned with the reception and correlation of sensory information. They are the sound engineers of consciousness. Our prefrontal cortex is the music conductor: it orchestrates our thoughts and actions in accordance with internal goals. The orchestra of the mind begins to play together.

The experience of conscious awareness is a global state event empowered by a dense network of specialist cortical neurons with long-distance connections that link the prefrontal cortex with other associative cortexes. These neurons are the biggest cells in our body, with their axons – the 'connecting cables' between neurons – extending several centimetres in length. How and when they begin to fire plays a significant role in our becoming conscious. When a stimulus enters the brain it upturns the ongoing fluctuations by reducing them or shifting them, and imposes new frequencies of its own. Using EEG, Dehaene measured electrical activity in the brain at the moment of perception, and identified a second signature of consciousness: a global high-frequency oscillation he called the 'P3 wave'.[13] This electrical wave acts like a positive feedback loop that amplifies the brain activation. By tracking brain activity with electrodes placed deep inside the brain, two more signatures were observed: a late and sudden burst of high-frequency oscillations and a synchronisation of information exchanges across distant brain regions. Synchronisation seems to facilitate transmission of information. Like a choir singing louder than a single voice could, information disperses quickly across the brain and we become aware of it. According to Dehaene: 'A conscious state starts approximately 300 milliseconds after the stimuli onset. The frontal regions of the brain are informed of sensory inputs in a bottom up manner. These regions send massive projections in the converse direction, top down, and to many distributed areas. The end result is a web of synchronised areas that provide signatures of consciousness: distributed activation, a P3 wave, gamma band amplification, and massive long distance synchrony.'[14]

Understanding and manipulating consciousness is of great medical and social interest. As populations in the developed world age, neurodegenerative diseases such as Alzheimer's and Parkinson's are becoming increasingly prevalent. The costs of health and social care are expected to shoot through the roof in the ensuing decades of the twenty-first century. Older people costing more money to support for a longer time means bulging national deficits and higher taxes for a shrinking population of young workers – a tearing-up of the social contract between generations and a recipe for social instability. In anticipation of this, billions of dollars are currently being invested by pharmaceutical companies and governments in brain research. Thanks to technological breakthroughs in brain imaging, and socio-economic pressure to deal with brain diseases, the human brain now attracts the same political attention as thermonuclear energy, fundamental physics and space exploration. This means publically funded scientific collaboration on a grand scale along the lines of organisations such as CERN,[15] ISS[16] or ITER,[17] where many countries and institutions come together under the same roof in order to invest talent and money in solving long-term, complex scientific problems and establishing infrastructures. 'Big science' is attacking the problem of brain disease, and – apropos – of consciousness.

In April 2013, President Barack Obama announced the BRAIN[18] Initiative, a collaborative program modelled after the Human Genome Project. Although the exact research agenda remains vague at the time of writing, a number of interesting ideas are being discussed. These ideas range from novel measurement methods of neuron activity using nanoprobes, to large-scale computation of models that simulate neural connectivity. But the most ambitious project on the brain today is undoubtedly the 'Human Brain Project' (HBP). Funded by the European Union and directed by the Ecole Polytechnique Fédérale de Lausanne (EPFL), HBP aims to build a complete computer model of a functioning brain. It is a truly global project involving hundreds of researchers and dozens of institutions from thirteen countries.[19]

The goal of the HBP is no less than the design and construction of an artificial human brain. Much research has already taken place in this area. For instance, anatomical maps of the brain have been the objective of a decade-long research project funded by the Allen Institute of Brain Science in Seattle, in the USA. Scientists at the Institute have managed to map the mouse 'connectome', the first brain-wide neuronal connectivity map for a mammalian species. This is a valuable piece of research that will guide researchers at the HBP, as they attempt to model the connectivity of the human brain. Nevertheless, the challenges of collecting and collating disparate information from previous brain research, and beginning to build models or neurons and groups of neurons in a computer, are simply huge.

The operational challenges of HBP, in terms of processing power, memory storage and just cooling the computers, are also enormous. The human brain is made up of eighty-nine billion neurons that connect with each other through hundreds of trillions of synapses – and all these computational elements must be coded. It will be the ultimate bottom-up exercise in reductionist science: to model a functioning brain using the fundamental elements of neurobiology – the neurons and their connections. If successful, the HBP will have tremendous impact in drug discovery. Experimenting with new brain drugs in a computer is much faster than doing the same in a wet lab, or in time-consuming high-risk clinical trials with human subjects.

Of course, apart from the obvious benefits to medical research, the HBP poses a deeper, philosophical question: will the digital simulation of the human brain be conscious? Assuming the simulation is built around a 'person', will the simulation have a personality, too? Including internal goals, needs and aspirations? Will it talk to us, and report its dreams, fears and hopes? If outcomes such as these were to take place, the scientists at HBP will have achieved an amazing feat. They will have furnished with intellect a gigantic mass of silicon chips, optical fibres and copper wires. They will have animated dead matter, and history will have taken an ironic, gothic, very Hoffmannesque twist: for the Human Brain Project is housed on the northern bank of Lake Geneva, only a few miles from the Villa Diodati where Mary Shelly penned *Frankenstein*.

Towards a theory of consciousness

Whether the artificial brain of the Human Brain Project will be truly conscious or not is a question that only a theory of consciousness can answer. But what exactly is a theory? Isn't it enough that talented scientists like Dehaene have unveiled the neural signatures of consciousness?

In science, a theory is defined as *a verifiable (and falsifiable) explanation* of facts and observation. It is the answer to a deeper 'why' question. More than that, a scientific theory can predict phenomena not yet observed. It is therefore not enough to know and measure what happens in the brain as information transits from non-conscious to conscious. We need to explain *why* it does so as well. Let's take an example from physics. People have been observing the stars for millennia. Detailed maps of astronomical observations have been found in ancient Mesopotamia, ancient Egypt and pre-Colombian America. Very probably these astronomical data have been collected and analysed systematically since the dawn of the agricultural revolution. But it took ten more millennia until Newton, and then Einstein, developed the scientific theory of gravity that explains *why* the celestial bodies move the way they do. Moreover, gravity theory (or the 'General Theory of Relativity' as it is formally called) explains much more than the planetary motions. It explains the whole cosmos. And it has predicted the existence of black holes, dark matter and dark energy. Similarly, a scientific theory of consciousness must not only explain why the brain achieves consciousness the way it does, but provide predictions of other phenomena relating to consciousness, for example dreams, hallucinations, consciousness in animals, schizophrenia, locked-in-syndrome, and others. Ultimately, we will need to know how it feels to be a bat, or how it feels to be you.

Stanislas Dehaene and his collaborator Jean-Pierre Changeux have generalised their results in order to offer such a theory. The Dehaene–Changeux theory of consciousness assumes a limited-capacity 'global neuronal workspace' in the brain. There, information – such as images, memories and feelings – comes together and is broadcasted to all brain regions through giant neural cells whose long axons criss-cross the cortex. Consciousness emerges in this workspace. The Dehaene–Changeux theory has many resemblances to the so-called 'blackboard' architecture of Artificial Intelligence systems. In such systems, a common knowledge base is constantly updated by a diverse group of knowledge sources until a solution to a given problem is found. The solution – the 'aha!' moment of the computer – seems similar to the moment of awareness in our brain that occurs when multiple and diverse, non-conscious processing of specific sets of information comes together into a whole. The 'whole' in the brain's case is driven sometimes by our basic survival goals and often by ephemeral desires or wants. These goals, desires, wants are 'problems' that our brain aims to solve by using its long axon hardware and 'global neural workspace'.

The Dehaene–Changeux theory was received with much applause, as well as criticism. Christof Koch, Francis Crick's collaborator and currently the Director of the Allen Institute for Brain Science, has claimed that the theory does not really explain the 'why', and that it is too focused on waves of electrical activity in the brain. Quoting the Russian-born novelist and scientist Vladimir Nabokov, Koch notes that 'the breaking of the wave cannot explain the whole sea'. Koch is fonder of an alternative theory proposed by Italian neuroscientist Giulio Tononi, currently at the University of Wisconsin. Tononi is an expert in sleep medicine, and has collaborated with Nobel laureate Gerald Edelman in consciousness studies.[20] His 'integrated information theory' of consciousness[21] aims to reinterpret Shannon's information theory in the context of awareness. For Tononi there is a profound link between information and consciousness. He defines 'Φ' (phi) as a measure of the complexity of an information system made up of causally interconnected parts. This measure is high if the system is extremely integrated and exhibits functionality that is not present in its parts, and also when the system has a large repertoire of discriminable states. The more integrated the system, the higher the Φ, and the more conscious it is. The human brain is obviously such an information system. Tononi's theory provides a probabilistic and testable framework for consciousness that could predict a value for Φ in animal brains, computers or even plants. Undoubtedly, this is one of its most valuable and appealing features. In fact, Koch believes that '*. . . in the fullness of time, the quantitative framework outlined in Phi will prove to be correct*'.[22] These are strong words coming from one of the most celebrated researchers in consciousness today.

Empiricism compels us to assume a deeper, material connection between information and consciousness. And yet, we are still quite far from reaching a scientific theory for consciousness that can satisfy us with its explanatory power. Although there has been incredible progress in the study of conscious phenomena, many questions are left outstanding. We do not yet know how millions of neuronal discharges distributed across time and space encode a conscious representation. In other words, we do not know if there is a direct correlation between specific patterns of neural discharges and whatever we become aware of at any given moment. We need new methodological tools and massive computer analysis of data to begin to understand how this happens. When we manage to understand this, we may then be able to answer questions such as why language exists, and how it came about. Could humans have evolved alternative ways of encoding experiences in the brain? Could we develop measuring instruments that will not only identify conscious experiences, but also communicate conscious experiences between different individuals, or even between different species? Could we explain synaesthesia, where perception follows different cognitive pathways? And what of the plethora of other neurological conditions?

If Φ is a measure of consciousness, it is probably a very coarse one. We still have much to explore and learn. The human brain is the product of four billion years of evolution, and we have not yet fully understood how evolution really works. And what about the body? What is it good for? Is it just an appendix to the brain, a crude, mammalian instrument for finding nourishment and sex? Does it play any role in consciousness?

Minds and bodies

Paradoxically, perhaps, many neuroscientists assume that, when it comes to thinking brains, the body is an unnecessary luxury. Their research implicitly subscribes to the notorious 'brain-in-a-vat' thought experiment, borrowed from mind philosophy. According to this thought experiment, a brain can be fully conscious without a body. All you need to do is place the brain in a vat full of nutrients and make sure it gets enough oxygen so that it does not die. Next, you connect the brain's neurons to a supercomputer via electrodes. The supercomputer can replace the real world with an artificial one. It can artificially construct illusions where electrical simulations of images, colours, smells and sounds can be transmitted via the electrodes to the brain. If that brain were yours – so the thought experiment goes – you would not be able to tell the difference between the real and the unreal. Like the hapless fellows used for batteries by the evil machines in *The Matrix*, you would be dreaming your life away while thinking it was the real thing.

In studying the brain as an autonomous system, without taking into consideration the rest of the body, neuroscientists have implicitly accepted that brains, and minds, are unitary entities in themselves. Nevertheless, biology seems to refute the brain-in-a-vat argument. Bodies of animals, including ours, are not made up of autonomous 'parts'. Yet we choose to regard them that way because this approach helps us understand anatomy more easily, and also because our medicine has evolved into studying and treating anatomical areas, systems and body parts, rather than the body as a whole. Could this fragmented concept of what a body is be preventing us from a deeper understanding of consciousness?

In the mid-1980s, the American neuroscientist Candace B. Pert (1946–2013) announced the discovery of a class of molecules in the body that she called 'neuropeptides'.[23] Peptides are molecules that intervene between the neurological, hormone and immune systems. They are short chains of amino acids that attach to special receptors existing on the surface of cells. In doing so, peptides communicate information between the various systems in our body, creating a vast psychosomatic network. Pert discovered a class of peptides that are also neurotransmitters, the endorphins. They resemble opiates, substances such as opium and morphine, and produce feelings of wellbeing as well as resistance to pain. Endorphins are the 'feel-good' agents in our brain, produced by the pituitary gland and the hypothalamus during exercise, love, sexual activity, pain and sometimes when we eat very spicy food. Pain and pleasure meet thanks to endorphins. Endorphins are also produced by the immune system,[24] a fact that enforces the hypothesis of a strong interconnection and bi-directional communication between the central nervous system and the rest of the body. It is therefore possible that the biological mechanism of consciousness is not localised in the brain but distributed throughout the body.

The connection between the immune system and the central nervous system points to another, perhaps less obvious aspect; that our bodies are also part of a larger whole that is the Earth's biota, i.e. the total collection of organisms that live on our planet. We become aware of information about the world mostly through our senses. Apart from the five traditional ones (smell, touch, vision, taste, hearing) we have several more:

balance and acceleration, temperature, a kinesthetic sense that helps us feel where the parts of our body are, pain, and several 'interoceptions' – senses of things that happen inside our body (for example, when we feel suffocation, or when we blush). Information coming from all these senses is processed by both our central and peripheral nervous systems. But the fact that our immune system connects to our nervous system suggests that we are also in communication with intruding viruses and bacteria. How might these interactions affect the way in which information finally reaches our conscious awareness? Can we dismiss our body's internal interactions and those with microorganisms as insignificant noise? As information that is never processed at the neuronal level? And what about interpersonal relations, the foundation for the survival of social apes like us? How does our consciousness modulate and transform from childhood to adulthood, and according to various social and economic circumstances? If our brain is a self-organising system made up of trillions of positive feedback loops that produce 'meaning' by amplifying information, then perhaps we may also consider how these recursive loops interact with our external environment and also with the rest of our body.

Perhaps a complete scientific theory of consciousness will have to wait for a paradigm shift in scientific thinking; one that will take us from reductionism to holism, and from the study of individual parts to the study of densely interconnected cybernetic systems. Perhaps neuroscientists ought sometimes to look towards cybernetics for interesting ideas, and possible answers.

11
THE CYBERNETIC BRAIN

Cybernetics is probably the most insightful, and ambitious, scientific synthesis of all time. By bringing together knowledge and paradigms from mathematics, physics, medicine and biology, the pioneering cyberneticians of the mid-twentieth century aimed to explore and understand the behaviour of complex, autonomous systems. But what exactly is a 'system'? A system can be broadly defined as a functioning collection of individual parts. The keyword here is 'functioning': the parts must somehow communicate with one another, and this exchange of information must guide the system towards a specific goal. Piles of sand or heaps of rubbish are not systems, although they are made up from many individual parts. However, a human with a pen and paper may be regarded as a 'system' for writing text. Pen and paper exchange information (via the brain connected to the hand that guides the pen) with the aim of committing a love letter, or a popular science book, to paper.

Cybernetics – from the Greek word for governor[1] – is interested in a particular class of systems that are also 'autonomous': once they come into being, these systems do not need their creator (or their 'first cause') any more in order to function. They can set their own goals and control themselves. The system pen–paper–brain–hand can be regarded as an autonomous control system. So a cybernetic system is also an 'automatic' system; it is self-controlled and self-guided. We could replace the human hand with a robotic one. And the human brain with an algorithm that composes sonnets, in which case we would have cybernetic poetry! We can actually think of numerous examples of automatic and autonomous systems. Think of the first automata in Hellenistic Alexandria, or a mechanical heart, or an autonomous robot exploring the surface of Mars.

But there also exist numerous other autonomous systems that are not products of engineering design. Arguably, every complex natural system is cybernetic. Think of the human body, or the nitrogen cycle in the atmosphere, or the fluctuation of algae populations in the oceans. Self-regulation seems to pervade natural processes at every level, from chemistry to cosmology. In the case of the human body and that of other warm-blooded animals, the cybernetic goal of self-regulation is called 'homeostasis'. During homeostasis the body system regulates its internal variables in response to external stimuli so that its internal condition remains stable. Take, for instance, human body temperature: our body achieves a more or less stable average temperature of about 37°C by means of multiple and interrelated feedback loops which receive information from the external environment and use this to regulate its biochemistry.

The word *homeostasis* is in fact very telling when it comes to how cybernetic systems self-regulate. The first part of the word, *homeo*, means 'by itself'; while the second part, *stasis*, means 'precarious equilibrium'.[2] A cybernetic system always tends towards a state of precarious equilibrium by constantly feeding back and integrating information about its environment and its internal variables. Such self-organised systems constantly hover at the edge of disorganisation, or chaos. Their states of precarious stability are often called 'attractors' because they tend to pull the system towards them. In the case of our body, life – the big attractor – is a constant battle against death.

Social and economic systems are also cybernetic. The 'invisible hand' of the market economy is the market itself acting as a self-regulating, dynamic, chaotic and autonomous system, made up of millions of individual and interacting parts whose goal it is to optimise the allocation of scarce resources. From a cybernetic perspective, the 'bust and boom' cycles in the economy can be viewed as catastrophic oscillations between various attractors. A good 'governor' of the economy would be the cybernetician who kept the market steady on the felicitous attractor of constant growth and plenty-for-all. If only it were that simple!

Cybernetics is therefore a 'meta-theory', meaning that it does not care what autonomous systems are made of, but how they function. Thus, cybernetics applies to any system, physical, technological, social, ecological or psychological – or any combination of these. Systems might connect to other systems and produce 'super-systems' that exhibit totally new behaviours. Combine the world's economy, its markets and societies with the Earth's carbon cycle, and you get a cybernetic super-system that affects the planet's climate in unpredictable new ways. Therefore, cybernetics provides us with a 'holistic' worldview of the cosmos. In contrast to reductionism and sciences divided into separate disciplines, cybernetics is emergentist and transdisciplinary. It seeks to understand emergent phenomena in which many autonomous systems, natural and artificial, interconnect and interact. This universal application of cybernetics was what Norbert Wiener aspired to. In this, he saw the opportunity to establish the ultimate science, one that would explain, and help direct, how systems can co-operate towards a desired goal. Taking this view to its logical conclusion, humans through purposeful actions could influence events beyond the confines of their planet, or their galaxy. Cybernetics could ultimately show us how to govern the evolution of life and the universe. At its inception, cybernetics was a very high stakes game indeed.

Nowadays, cybernetics seems to belong mostly to the history books. Although there are still a few research centres calling themselves cybernetic, the truth is that Wiener's science has acquired a steampunk aura. At best, it is remembered with nostalgia as the womb that spawned many modern scientific disciplines: computer science, artificial intelligence, control theory, information theory, cognitive science, computer modelling are but a few of cybernetics' surviving, and thriving, offspring. The eventual fragmentation of cybernetics speaks volumes about the socio-economic aspects of scientific research. Universities, research centres, governments all seem keen to resist holistic approaches to knowledge, preferring the departmental separation of sciences and humanities that keeps grants flowing and tenures going. Wiener's original dream of creating a science of everything seems shattered today. And yet the names, as well as the deeds, of the pioneering cyberneticians remain with us still: von Neumann's computer architectures, game theory and cellular automata; Ashby's and von Foerster's analysis of self-organisation; Braitenberg's autonomous robots; and McCulloch's artificial neural nets, perceptrons, and classifiers, still inspire researchers and students around the globe.

Understanding cybernetics is crucial to understanding Artificial Intelligence. The two fields are very closely related. Indeed, the human brain and questions about the mind were what instigated and catalysed the birth of cybernetics. It all started at the Macy Foundation for Public Health and the Medical Professions in New York, under the direction of one visionary, as the Second World War passed the baton of global annihilation to the Cold War.

The Macy Conferences

Frank Fremont-Smith (1895–1974) was an American administrator with a keen interest in the brain and the functioning of the human body.[3] He spent the 1930s studying the pioneering work of the American physiologist Walter Canon on homoeostasis. In the early 1940s, he started a discussion group that examined reflexes and hypnosis. Several pioneering scientists attended the group, including Gregory Bateson and Margaret Mead, and they called themselves the 'Man–Machine Project'. When Fremont-Smith became the

Medical Director of the Macy Foundation, he set up a series of annual conferences that expanded the Man–Machine Project. Hosted by the Macy Foundation, these became known as the 'Macy Conferences on Cybernetics'. Cybernetics as a field grew out of these interdisciplinary meetings, held from 1946 until 1953, which brought together a number of notable post-war intellectuals, including Norbert Wiener, John von Neumann, Warren McCulloch, Claude Shannon, Heinz von Foerster and W. Ross Ashby. From its original focus on machines and animals, cybernetics quickly broadened in scope to encompass the workings of the mind (e.g. in the work of Bateson and Ashby) as well as social systems (e.g. Stafford Beer's management cybernetics), thus rediscovering Plato's original focus on the control relations in society. I will return to the very interesting connection of cybernetics, Plato and global governance later in the book. For now, I want to focus on four individuals who took part in the Macy Conferences, and whose work laid the foundations for Artificial Intelligence: Norbert Wiener, Claude Shannon, Warren McCulloch and John von Neumann.

We have already met the first two. Norbert Wiener was the grand visionary of cybernetics. Inspired by mechanical control systems, such as artillery targeting and servomechanisms, as well as Claude Shannon's mathematical theory of communication and information, he articulated the theory of cybernetics in his landmark book, *Cybernetics*, of 1948.[4] Godfather number two, Claude Shannon, was the genius who gave us information theory. We saw how Wiener and Shannon pondered on the ontology of information, and how they decided to regard it as something beyond matter and energy. The legacy of their decision is still with us today, in the disembodiment of information that defines post-humanism.

The man who demonstrated the direct connection between neurons and computers was Professor Warren S. McCulloch (1898–1969), the American neurophysiologist who loved writing sonnets and laid the foundations of many contemporary brain theories. In 1943, he collaborated with Walter Pitts, a logician, on a seminal paper about the mathematics of neural cells.[5] In this paper, McCulloch and Pitts tried to understand how the brain could produce highly complex patterns by using many basic cells – called neurons – that are connected together. To do so they borrowed ideas from Alan Turing.

Turing's influence has been tremendous in America, and his ideas for calculating machines (the so-called 'Turing machines') provided an excellent theoretical framework for McCulloch and Pitts. In their paper, they demonstrated how neurons could be equivalent to programs run on a Turing machine. In doing so, they effectively proposed that neurons might be regarded as information processing machines, and as the base logic units of the brain. This one-to-one correspondence between neurons and programs has since become one of the central tenets of computational theory and Artificial Intelligence.

If the brain is made out of base logic units that process information, then intelligence ought to emerge from the interconnectedness between these units. The brain is therefore a cybernetic system. As Dehaene's research into consciousness has shown, the brain uses feedback loops that pass information from neuron to neuron and from groups of neurons to groups of neurons. Sensory inputs from the nervous system are continuously integrated at a neural level. These integrations affect internal states in the brain, such as memories and thoughts. Intelligence is an emergent behaviour as the brain instructs the body how to react and respond to external stimuli. If this hypothesis is true then the brain can be replicated in any medium that can process information in a similar, granular, logic unit base fashion. This medium could be made up of gears, nuts and pulleys, silicon chips, or water pipes – it does not matter what it is made of as long as it can process information in a similar manner to that of the brain.

The McCulloch and Pitts model of a neuron, often called

'*MCP neuron*' for short, made a seminal contribution to the development of artificial neural networks. The model also formulates the central problem of AI as corresponding directly to the problem of human consciousness. If we can discover how the cybernetic brain thinks, then we will have solved the problem of AI at the same stroke. As we shall see, this is exactly what one of the most important theories of Artificial Intelligence claims: Marvin Minsky's theory of intelligent agents. Like neurons, or groups of neurons, autonomous software programs called 'agents' compete with each other while trying to solve a problem. Sometimes they collaborate while other times they antagonise each other. The end result of these chaotic interactions is the amplification of collaborating agents that quickly disperses throughout the system. As the system self-organises, it goes through successive phase transitions, or 'bifurcations' – as the more technical term goes. At each one of these bifurcations new functionalities emerges. Finally, there comes a tipping point, where global change happens and the artificial, agent-based system becomes intelligent, in a similar fashion to the neuron-based human brain.[6]

The fourth cybernetician godfather of Artificial Intelligence, who also took part in the Macy Conferences, was the legendary Hungarian-American mathematician John von Neumann (1903–1957). He was the modern equivalent of Gottfried Leibniz, a polymath who made fundamental contributions to several sciences including mathematics, computing, cybernetics, logic, economics and quantum physics – to name but a few! His last work, before his untimely death at the age of fifty-three, was an unfinished manuscript entitled 'The Computer and the Brain', which shows how deeply interested von Neumann had become in the nascent science of Artificial Intelligence.[7]

During the time he participated in the Macy Conferences, von Neumann expanded on his theory of self-replicating automata. He demonstrated that machines could potentially not only think and behave like humans, but might have the capacity to reproduce as well.

Self-replicating machines

Von Neumann was familiar with Turing's work. In 1936, he wrote a reference for Turing, supporting his application for a Fellowship at Princeton. The two men had met a year earlier when von Neumann was visiting professor at Cambridge University. By 1939, von Neumann must have read Turing's work on automata and computing machines,[8] for he used his approach to study how cybernetic systems could self-replicate.

Turing had shown how a machine could code any kind of information – a concept he termed a 'Universal Machine'. Von Neumann realised that, in essence, this meant the Universal Turing Machine could also code itself. Indeed, modern computers, which are Universal Turing Machines, have exactly this ability. All software stored in your computer can be copied to another computer, by your computer. In fact, copying is what takes place whenever you perform any transaction using a computer. When you 'send' an email, for example, nothing actually moves from one place to another: an exact copy of your email is reproduced in the computer of the person you want to communicate with. Von Neumann was fascinated with this self-copying property of the Universal Turing Machine. In true cybernetic fashion, he set off to formulate a general theory of self-reproduction that would include living organisms as well as machines. He embarked on this quest with a series of lectures in 1948 – four years before Watson, Crick and Franklin discovered the molecular structure of DNA – and thus answered the puzzle of self-replication in living systems.

Von Neumann applied mathematical logic to show that there exists an automaton that can produce at least two copies of any description of another automaton you feed it with. To prove this theorem, he used a method known as 'logical substitution'. We will encounter this method again in the next part of the book, when I will discuss logic in more detail and how that Austrian mathematical prodigy, Kurt Gödel, used the method to prove logic's limits with his infamous 'Incompleteness Theorem'. As it happened, Turing also used logical substitution in order to prove his theorem on incalculability. Logical substitution is one of these magnificent tricks that mathematicians keep up their sleeves to get them out of trouble. It creates a one-to-one correspondence between two classes of mathematical objects, say between natural numbers and logical signs or operators.[9] Thus, the smart mathematician solves an easier problem rather than an impossible one, and becomes famous forevermore.

Using logical substitution, von Neumann substituted bits of information (for example, sets of positions on the infinite tape of a Universal Turing Machine, or a 'program' as we would call it today) with whole Turing machines, in order to prove his theorem for self-replicating automata. And here's the deep insight of this substitution: it confirms, in a most profound way, that artificial life can reproduce in exactly the same way as biological life. Any living system with the ability to replicate can do two things: it can produce and self-replicate. Take genes, for example – they code information for producing proteins but also code for their own self-replication as well. Long before genes were discovered, von Neumann had the insight to show that self-replication and production (or 'copying' in the sense of modern computers) are the twin properties of *any* self-replicating system, including self-replicating automata. He called his self-replicating automaton 'the Universal Constructor', to echo Turing's 'Universal Machine'.

Although von Neumann did not live long enough to complete his theory of self-reproducing automata,[10] the impact of his big idea has been tremendous. The discovery in 1953 of how DNA is structured, and the subsequent biological understanding of how life replicates and evolves, answered several centuries-old questions about life.[11] But the discovery of the DNA did not suffice to answer the most crucial one: *how did life come to be*? Interestingly, von Neumann's theory of self-replicating automata provides some very interesting answers to that question, so let's see what cybernetics can tell us about *real* life.

A Universal Constructor can construct anything that can be constructed. Nevertheless, Turing showed that you couldn't build a machine to tell you in advance whether something can be constructed or not. A Universal Constructor can only replicate whatever information you feed it with. It cannot invent itself. If we accept that living systems are Universal Constructors and go back to the very beginning of evolutionary time, we must ask ourselves what information was fed (and by 'whom') to the very first constructor that set off the creation of life. It is the same question as asking, more simply, '*Why* is life possible?'

This question can be answered in only two ways. You could assume the presence of a programmer, i.e. a Creator, or an agency that 'knew' *a priori* that life could exist, and which fed this information into the primal molecular automata that would evolve over time into multicellular, conscious organisms. Let's call this explanation 'argument from design'. It appears to be logical, and several billion people on our planet believe this argument in earnest by adhering to apocalyptic religions such as Christianity and Islam.

Nevertheless, there is a very serious logical problem with this argument called infinite regression: who or what created the Creator? If it was another Creator, then we regress to the same question, again and again, ad infinitum. There may be infinite universes and infinite Creators of universes, but we must still wonder how the 'first' Creator came into being. How could it be known in advance by any self-replicating constructor that life was possible? In the absence of DNA molecules or other physical memory, where was the ur-information for life encoded and stored?

We simply cannot answer this. Unless of course we accept a Platonic perspective and place blind faith in universal forms that exist outside a material universe. Unless we believe that there is a mathematical blueprint for carbon-based life in the metaphysical archives of universal forms.

However, if, like me, you are not quite content with Platonic metaphysics, then we must consider another explanation for life: that complex automata (e.g. bacteria, animals, humans) can arise from very simple automata (e.g. autocatalytic chemical reactions). We must conjecture that complexity must arise from simplicity, and that the wonderful intricacy we observe in nature, from the very small to the very large, in structure as well as in function, has extremely humble and simple origins. Once we accept this then our question about the origins of life – and intelligence – comes down to understanding not *whether* but *how* the complex emerges from the simple.

This scientific, and essentially cybernetic, question of emergence is directly relevant to Artificial Intelligence. In aiming to make machines think and feel like biological humans, we must somehow reconstruct the complexity of the human brain from simple computational units. Intelligence must somehow emerge from the interactions between these elementary computational units, those non-conscious agents, just like it emerges in the brain from the elementary non-conscious neurons.

As we shall see in the next part of this book, it is not possible to code the whole spectrum of human intelligence. There are functions of the human mind, notably intuition and induction, which lie beyond logic.[12] As we shall also see, Gödel and Turing showed that we must distinguish between the purely logical (which can be coded) from the intuitive and the inductive (which cannot). The ultimate goal of Artificial Intelligence to engineer a conscious machine can therefore be realised only by following a cybernetic approach: i.e. to allow complexity to emerge spontaneously from simplicity. This emergentist idea has been tried in at least two of the typical 'AI programming languages' – Prolog and LISP, which follow an iterative approach in their syntax. A programmer begins by defining simple local rules and then, through highly recursive structures (or complex feedback loops), allows complexity to emerge spontaneously. To understand how this can happen, we need to examine two key cybernetic ideas in more depth: reflexivity and the emergence of order out of chaos.

Reflexivity and the emergence of order

The two-in-one property of the Universal Constructor suggests that when a system generates another system (whether it is an exact replica of itself or another product) it becomes part of what is generated. This process is called 'reflexivity'. Reflexivity is the moment whereby that which has been used to generate a system is made, through a changed perspective, to become part of the system it generates.[13] Logical substitution – the method that von Neumann used to prove that a Universal Constructor automaton exists – is applied reflexivity in mathematics. Reflexivity is the cybernetic concept of feedback taken to the next level: the circular relationship between cause and effect, self-referencing without end. Nevertheless, this is not an abstract or impossible-to-visualise concept. Imagine a video camera in front of a mirror recording itself: the result is what reflexivity *looks like*. Reflexivity is very much an Aristotelian, objective quality of the physical world. It can be seen and measured. It is as real as anything.

In fact, reflexivity may also be the prime cause of life. There is a class of chemical reactions called autocatalytic that are a chemical demonstration of reflexivity. In autocatalysis, the product of the reaction is also the catalyst of the reaction. Reactions of this kind play a vital role in biological systems, from how rRNA is transcribed to how haemoglobin binds oxygen in the blood. Metabolism is in fact a vast autocatalytic reaction since all the molecular constituents of a cell are produced by reactions involving the same set of molecules. Autocatalysis may be how life first began, and how the transition from the simple to the complex took place and evolved. *Abiogenesis* is a theory that postulates that life arose as autocatalytic chemical networks. Experiments have shown that certain autocatalysts react to environmental changes, and are thus susceptible to natural selection.[14] Reflexivity, an intrinsic property of cybernetic systems, may be the key to life.

Reflexivity can also be felt – just think of the words 'I am'. You think of you, who thinks of you, who . . . That's how reflexivity *feels like*. When reflecting upon myself, I am object and subject at the same time. Psychologists call this recursive sense of consciousness 'metacognition'. The observed and the observing are encoded at different times or within different systems. It is only when this self-referencing ceases that the self also ceases to exist. This is a very significant conclusion. It tells us that reflexivity, an idea at the core of cybernetics and an essential property of self-organising complex systems, may also hold the key to engineering machines with consciousness.

Reflexivity guides the emergence of order because it conditions the positive feedback loops in a cybernetic system. As we saw, one of the most frequent criticisms of emergentism is that it is 'mysterious', and resembles vitalism, because it does not suggest a convincing chain of cause and effect like reductionism does. The counterargument is that cybernetics is an empirical science grounded in physicalism, the theory that the universe is made up exclusively of physical entities. It does not assume the existence of mysterious forces from outside the material universe. Instead, complex systems are studied through the relationships between their fundamental blocks, and therefore the principles of reductionism are not completely abandoned. On the other hand, pure reductionism is insufficient to explain the emergence of new properties in complex systems. To provide a scientific explanation of an emergent property –for example, consciousness in the brain – we need to follow a cybernetic approach. We need to define successive levels of self-organisation: from molecules, to neurons, to anatomic brain areas, and perhaps all the way to society at large – and study the cause-and-effect chain of successive bifurcations that lead to the emergence of consciousness.[15] The findings of neuroscience and modern research in consciousness seem to confirm that a cybernetic approach is the proper one. As Dehaene and others have shown, all our senses are ambiguous. Take, for instance, vision. Our brain receives a visual signal through the optic nerve – an elliptical shape, say – and begins to process it. During the intermediate stages of processing the visual signal, the brain ponders a vast number of alternative interpretations of its sensory input. A single neuron may perceive only a small segment of the ellipse's contour. Once neurons start talking to each other and casting their votes for the best fit, the entire population of neurons converges. Convergence follows the logic of Sherlock Holmes: eliminating the impossible leaves the truth, however improbable. Rephrasing this sentence in mathematical terms, our brain performs reverse referencing every time we reason from outcomes to causes. It is like seeing footprints in the sand and deducing from them that another person must have walked there before us. This is statistical reasoning in a backward manner – to infer the hidden causes behind observations. The brain considers all possible causes, weighs the relative evidence against them, and decides on the most probable cause. This suggests that there is a deep connection between mental processes and statistical computation. It also suggests that our brain has all the characteristics of a 'second-order cybernetic system'. Let's see what that means.

In first-order cybernetic systems, emergence is a property observed by an outside observer. Under this heading, we can classify the vast majority of complex systems in nature, for instance migrating flocks of birds and weather systems, but also engineered systems such as automated manufacturing processes and control systems. In second-order cybernetics the observer is an integral part of the system that is observed.[16] The epistemological dichotomy between object (what is observed) and subject (who does the observing) breaks down. Second-order emergence in the brain is when neurons connect and give rise to memories, sensations and thoughts that, in turn, observe the brain itself. It is this type of self-referential emergence that is most interesting to Artificial Intelligence. If we can make artificial systems evolve second-order emergence then we will have engineered artificial consciousness. And the key to understanding how second-order emergence arises in living systems is reflexivity.

The fugue of the mind

In 1979 the American mathematician and philosopher Douglas Hofstadter published a ground-breaking book[17] that explored how self-reference and formal rules allow meaning to emerge from meaningless elements. The book created a sensation because, apart from its very serious scientific premise, it was also inspired by art. Entitled *Gödel, Escher, Bach: An Eternal Golden Braid*, the book used narratives, paradoxes and logical arguments to explore the connection between the Austrian mathematician who discovered the limits of logic, the Dutch graphic artist who challenged our visual perception and the German composer who produced some of the most beautiful music ever. All three used self-referencing, or reflexivity, in their work.

I have mentioned in passing how Kurt Gödel used logical substitution to prove his incompleteness theorem, and thus prove that logic is not sufficient to prove that something is true. We will discuss this earth-shaking theorem in more detail in the third part of this book. For now, let us marvel at what Gödel did in order to arrive at his proof. By applying the principle of 'mathematical reflexivity', Gödel's proof used the proof itself in order to arrive at the truth of his argument. I have always found his stroke of genius akin to the pleasure I get when I listen to beautiful classical music. There is an aesthetic quality in mathematical truths, as many mathematicians would testify. Truth is beautiful, and beauty attracts us. Music can be beautiful, too, and the connection between mathematics, music and aesthetics is something that you don't have to be a Pythagorean in order to appreciate. Take, for instance, the musical genius of Johann Sebastian Bach (1685–1750). He has bequeathed a rich legacy to our musical culture, including some of the most enthralling fugues ever written. Bach was the master of the fugue, and fugue is a particular musical composition that uses the so-called 'canon form'. This is a technique in which a composition consists of a leading melody (called the *dux*), and several other imitations of the leading melody that are played with a higher or lower voice after certain duration.

The fugue is a musical example of self-referencing that creates aesthetic beauty. But isn't it the musician's mind – in this case Bach's – that created this form of self-referencing before committing musical notes to paper? And isn't it our listening mind that feels the pleasure as the instruments play the score? Seen this way, music is the *mediator* between the self-referencing in Bach's mind and the self-referencing in our minds. Every time we listen to one of Bach's fugues we *become the music score*.

M. C. Escher (1898–1972), the third name in Hofstadter's book title, was deeply influenced and inspired by mathematics. In his mind-bending work, Escher depicted impossible constructions based on multi-dimensional geometry, and explored the meaning of infinity. One of his best-known lithographs is of two hands drawing each other: the object becomes the subject, which becomes the object, in a never-ending cycle. It is exactly this recursive relationship between the objective and the subjective that Hofstadter calls our attention to. Recall the example of the video camera recording itself. The observer (the camera) is the subject; the observed (the camera) is the object. Through self-referencing, subject and object become the same. The result of this union – the infinite reflection of the camera-within-the-camera – is Hofstadter's 'golden braid', the infinite interlacing of subject becoming object, from which meaning emerges.

But what exactly is 'meaning'? From a cybernetic point of view, meaning is a new functionality, something that the parts of the system do not – and could not – possess. One neuron alone can find no meaning in a fugue. But a collection of neurons firing excitedly across all regions of the cortex can. Hofstadter claimed that cognition and thinking emerge from neurological mechanisms in the brain through complex positive feedback loops. The human brain can thus be viewed as a self-referencing biological entity that attains consciousness through self-referencing. The human mind *is* a 'strange loop' – as Hofstadter likes to call it.[18] Descartes would probably have agreed. His definition of consciousness – '*I think therefore I am*' – is a fine example of recursive thought: the mind exists because it can think itself. We bootstrap ourselves in existence every time, out of nothing, by self-reflection.

We saw in Part I how theory of mind is one of the most fundamental cognitive characteristics of the typical human mind. Theory of mind is also recursive. It gives us the ability to conceive our minds and the minds of others. It also allows us to perform mental time travel: thanks to theory of mind we can bring into our present consciousness events that took place in the past, as well as imagine ourselves in various, hypothetical future situations. The evolutionary advantages of recursive thinking are enormous. A cognitive system capable of meaning can strategise more effectively about future events and eventualities. This strategising is the essence of free will. Although our choices are never infinite, owing to circumstance and how physical reality is arranged at any given moment, we are always free to consciously decide and choose thanks to our ability to self-reference. Recursive thinking can also maximise the benefits of social living by manipulating other minds. It is very probable that recursive thinking is what begot general-purpose language[19] in *Homo sapiens*. After all, our language constantly uses recursive syntax to communicate recursive thoughts, as Descartes' famous dictum demonstrates. Our embodied brains are second-order cybernetic systems that produce an emergent property vital for their survival, which we call consciousness.

So, what is the mind?

I started this part of the book by asking what the mind is, and then searched for an answer in philosophy, neuroscience and cybernetics. It was not easy to arrive at a definition of the mind that would be acceptable by general consensus, because of a deep, historical dichotomy in Western philosophical thinking between form and matter. This dichotomy prevents us from reaching a scientific understanding of the human mind, because we cannot clearly decide whether form takes prevalence over matter, or if it is the other way round. Unfortunately, we remain under the influence of the opposing worldviews of two ancient Greek philosophers, Plato and Aristotle.

Plato's influence has been enormous, probably because it resonates so perfectly with the dualist nature of our cognition. In other words, we are all born Platonists. Resurrected by Descartes in the eighteenth century, Platonic dualism has become so deeply ingrained in Western philosophy and science that it sets the agenda for contemporary discourse on Artificial Intelligence and consciousness today. It was Plato who influenced the historical decision of the two cybernetic giants, Wiener and Shannon, to frame the nature of information as something distinct from energy or matter. This distinction has contributed to the disembodiment of information, and the false separation of the physical substrate (the hardware, the brain) from information patterns (the software, the self). Information thus became prevalent, the master of everything. Without software, hardware is useless.

Without consciousness a brain is comatose. Without DNA chemistry is just a soup. This 'computer metaphor' for life and consciousness, where the form, or pattern, takes precedence over matter, defines our post-human present, and justifies paradoxical predictions about downloading consciousness in computers and achieving digital immortality.

At the opposite side of the form-versus-matter debate, empiricists – the 'Aristotelians' – react vehemently to Platonism and dualism. For them, universal forms are always instantiated in physical objects. Knowledge does not exist *a priori*, but is the result of discovery through observation of natural objects, processes and phenomena. For Aristotelians, mathematics is *always* predicated in physical objects and never vice versa. They agree that there seems to be a deep connection between computation, life and consciousness. But for them, to discover this connection one must avoid the sirens of metaphysical, Platonic universals and explore nature. For empiricists, the problem of mind boils down to understanding how many non-conscious logical parts (called neurons) come together to produce a conscious whole (the conscious brain) – and there should be nothing magical, or mysterious, or Platonic, about that. However, the dualists retort, even if the empiricists answered that scientific question there would still remain the 'hard problem of consciousness', namely the inexplicability of subjective experience (also called 'qualia'). Eccles, Penrose and Hameroff suggested explanations of subjective experiences based on quantum physics; however, these explanations are either openly dualistic (Eccles) by postulating immaterial psychons, or unashamedly Platonic (Penrose and Hameroff) by suggesting that consciousness is interwoven in the geometrical, quantum space–time fabric of the universe.

And this is where cybernetics comes into play, in defence of empirical physicalism. Cybernetics claims that subjective experiences *can* be explained through self-referencing, or reflexivity, in the second order cybernetic system called 'brain'. One does not need psychons or quantum physics to explain self-awareness. Biology suffices. The hard problem becomes an easy one, and neuroscientists studying how the brain evokes consciousness appear to vindicate a cybernetic explanation. As Dehaene's research shows, the feeling of selfhood (a subjective experience) seems to emerge from non-conscious, self-referencing neurological processes. This is, of course, a scientific hypothesis not yet proven. However, huge strides have been made towards its confirmation thanks to empirical science, as well as countless days and nights spent observing and collecting data in neuroscience labs. We should expect that, as more sophisticated imaging instruments come to the fore, together with more computing power that can analyse massive experimental data, empirical science will explain consciousness in an Aristotelian, physicalist way.

Once we understand how nature 'does' consciousness, it will become possible to engineer a fully conscious artificial machine. However, as we saw, we will have to revise the brain-in-a-vat paradigm, which is adopted implicitly by neuroscience. We will have to reverse the trend that first disembodied the brain, and then disembodied the mind. Our brains are part of our bodies, and examining them as isolated objects misleads us. We are who we are because of how our body interacts with our physical and social environments, and how it has done ever since we were in our mother's womb. Our minds develop and constantly change because they are embodied. By the same token, conscious machines must also be embedded in a physical environment, and must have some kind of body that allows them locomotion, sensing external and internal stimuli, thereby enabling them to learn from their experiences through exploration.

Time, then, to leave philosophers and neuroscientists behind and enter the computer lab, the place where the dreams of artificial minds will soon become real.

PART III

ADA IN WONDERLAND

DAVE BOWMAN: Open the pod bay doors, HAL.
HAL: I'm sorry, Dave. I'm afraid I can't do that.

Arthur C. Clarke, *2001: A Space Odyssey*

12

'ALL CRETANS ARE LIARS'

Computers owe their existence to the curious idea that thoughts can be expressed as a collection of abstract rules. Furthermore, that these rules can be encoded using symbols, into what nowadays we call 'programming languages'. The first to explore the abstraction of thoughts was Aristotle. He suggested that thoughts ought to be the object of scientific curiosity and investigation, just like other phenomena in nature. Aristotle called the mechanism of thinking 'logic'.[1]

Following Aristotle's pioneering work, logic became the instrument with which scientists and scholars explored everything, from mathematics to physics, from economics to politics – all the way to the existence of God. With time, logic transformed from object to subject: like someone looking at a mirror, logic was used to explore itself. Culminating in the middle of the twentieth century, this transformation kindled new and very interesting questions. Could logical thinking be an empirical process? Is logic predicated on physical neuronal processes in the brain? Or is logic an ideal form, an immaterial scaffold of fundamental rules upon which the material universe is built? In both science and the humanities, the creative dichotomy between the Platonic and Aristotelian worldviews continues to imbue discussions about the nature – or ontology – of logic.

But let's unfurl the history of logic by starting with Aristotle, the father of logic. He was the first to realise that one could isolate the act of reasoning from the objects of reasoning. He called pure, distilled reasoning processes 'syllogisms'. A syllogism is therefore a series of thoughts, one following the other by necessity. For example, if A equals B and B equals C, then – necessarily – A equals C. Syllogisms can be used to deduce whether something is true or not. You start by supposing something is true, and then use a syllogism to conclude whether your supposition is actually true or not. Thus, the logical process of deduction creates new knowledge. Uncertain things become certain using syllogisms. Aristotle claimed that new knowledge can only be obtained through a combination of observation and logic; and this combination, as we saw in the previous part of the book, is the foundation of empiricism. Modern empirical science codified syllogisms in the scientific method, and continues to produce new knowledge about the world through using observations in conjunction with logic.

Aristotle also showed that there are three ways to combine observation with logic. The most effective method is called 'deduction'. There is no escape from a deductive process. Given a premise a deductive conclusion must necessarily follow. If A equals B and B equals C, then A equals C – and that's that! Nevertheless, scientific theories are mostly based on a weaker form of logic called 'induction', whereby one derives a reliable generalisation from observed facts. That's what scientists usually do when they experiment in their labs. Their experimental data are used to induce a general conclusion about a natural phenomenon. This inductive conclusion is usually called 'scientific theory'.

There are times, however, when it becomes difficult for scientists to use induction. This happens in cases where experimental evidence is not directly related to a studied phenomenon. Take, for instance, black holes. They cannot really be 'seen' directly. Nevertheless, scientists can make indirect measurements of other phenomena – for example, changes in gravity around a black hole – and draw scientific conclusions about the existence, and nature, of black holes, i.e. of something they could never measure directly. This method is called 'abduction': you start from some observation and try to guess the reason that could explain the relevant evidence. 'Abduction' is how a police investigator would solve a crime. Imagine a dead body found on a beach, and strange footprints next to the body leading away into the darkness. An investigator would study the footprints, and abduct a hypothesis about what that person – a possible suspect – might look like, or where he or she went after the crime.

Computers use all three logical methods in order to function. They take 'inputs' which are data, or observations about something. Then, they apply a logical process – deduction, induction or abduction – to produce an output, or a conclusion.

The Laws of Thought

Aristotle's logic dominated Western thought for centuries. The nineteenth-century British mathematician and logician George Boole (1815–1864) expanded on Aristotelian logic in his influential book *The Laws of Thought*.[2] In the book, Boole showed that one could use algebraic equations to express, and manipulate, logical sentences. Boole managed to translate English sentences into the language of logic using notation and ideas from algebra and mathematics. This 'algebra of logic' is nowadays called symbolic logic. By effectively reducing logic to algebra Boole demonstrated how the manipulation of symbols could provide a fail-safe method of logical deduction.

Here's an example of Boole's great innovation. By borrowing from algebra, Boole defined three basic logical operations: conjunction, disjunction and negation.[3] By conjunction he meant when two things are taken together in order to arrive at a conclusion. He denoted this relationship with the symbol $\wedge$ (which reads 'AND'). Boole defined this logical operation between two things – let's call them 'x' and 'y' since we are borrowing from algebra – as follows:

$x \wedge y = 1$ if $x=y=1$ otherwise $x \wedge y = 0$

Translating in plain English the above says: if x and y are the same, then when they come together we will call their conjunction '1'. If, however, they are different, then we will call their conjunction '0'.

Boole showed that by following this kind of symbolic notation one could build a very sophisticated system that processed logical statements. He was the first to connect logic to mathematics; and these two fields would remain intertwined forever after. By using the binary values '1' and '0' to denote 'true' and 'not true' respectively, Boole also laid the foundations of binary logic. This binary, 'Boolean logic' would become the basis of the modern digital computer and electronics.

Boole's symbolic logic was extended further by Gottlob Frege (1848–1925). Until the end of the nineteenth century, logic dealt with constants such as 'and', 'or', 'if', 'then', but found it hard – and sometimes impossible – to deal with constants such as 'some' or 'all'. For example, the sentence 'there are infinite prime numbers'[4] was impossible to denote with the existing notation. Frege invented 'predicate logic' in order to tackle such logical statements. Predicate logic contains, and manipulates, statements with variables that can be quantified. To follow with our example, Frege found a way to express the idea that, if you are a prime number then you belong to a class that has other infinite numbers like you. Therefore, you the prime number, are 'predicated', or affirmed, by belonging to this class.

In so doing, Frege had to invent a way to depict these 'quantifiable variables'. Thankfully, he was a formidable innovator and inventor of practical tools that dug deep into logic. Kurt Gödel and Bertrand Russell used Frege's notation and many other tools developed by him. Let's look at the two quantified variables that Frege invented, and use them in order to gain a deeper insight into this wonderful thing called predicate logic. The first is called 'universal quantifier'; it reads 'for all', and looks like an inverted 'A': ∀. The other is called 'existential quantifier'; it reads, 'there exists', and looks like a mirrored 'E': $\exists$. Let's see now how we can use these quantifiers to denote a sentence such as 'all Greeks are philosophers or like to drink coffee' – and then play around a bit!

First, let's denote Greeks with the capital letter X. Let us also denote any human being with the lower-case letter x. If you are a Greek then you belong to the group of Greeks. We can denote this simple statement like this:

$x \in X$ (where the symbol $\in$ means 'belongs to').

Greeks, according to our original sentence, are 'predicated' by two things. They are philosophers or like to drink coffee. Frege would encode these two 'predicates' about anyone (the guy next door we called 'x') using the following notation:

$P(x)$: means x is a philosopher

and

$Q(x)$: means x likes to drink coffee.

So, if you are a Greek then you are someone who $P(x) \vee Q(x)$. The word – or 'constant' as it is called in logic – 'OR' is denoted by the symbol $\vee$.

Let's now put everything together, and 'translate' the English language sentence 'all Greeks are philosophers or like to drink coffee' using logical symbols:

$\forall x \in X, P(x) \vee Q(x)$

Easy!

Using his notation, Frege developed a very sophisticated system of analysing logical truths. And then he went further. With his predicate logic and notification he aspired to develop a rigorous system that was not subject to the nuances and confusions of normal language, and that could be used to safely and unambiguously test whether any statement was true or not. He was aware of the problem of intuition in mathematics and science. Intuition is when we know something to be true but we can never hope to prove it. Frege realised that intuition could be represented as sets of axioms, unprovable truths that are fundamental for a formal system to exist. He showed that, once you excluded intuition from a formal logical system, then proofs about truths within that system had to be logical within that group. In effect, this notion illustrated that all mathematical truths are logical. It was a conclusion that, as we saw in a previous chapter, impressed and influenced the neopositivists of the Vienna Circle. Frege had effectively shown that there was a way to bypass natural, everyday language and use pure symbolic logic to think about everything. Predicate logic was indeed the royal, the divine, road to the absolute Truth!

Frege also showed that arithmetic is part of logic. Unlike geometry, where intuition plays a role and one has to accept axioms as true without proof,[5] arithmetic has no need for intuition and therefore no need for non-logical axioms. By proving this, Frege provided even more intellectual glue that bonded logic and mathematics together. By the early twentieth century, mathematicians and logicians were becoming one and the same. From their union the idea of a formal logical system was born. It would dominate the discourse on logic for the next forty years, as some of the most brilliant minds ever tried to decode the ways in which logic dictated everything.

Let's consider what exactly a formal logical system is. The word 'formal' implies the Aristotelian concept of 'form', whether it is applied to the human body or a plant, an animal, a river, a temple for Zeus, indeed any animate or inanimate thing. A system of logic can be seen as an autonomous physical entity. Therefore, like the human body, an autonomous, self-contained logical system also possesses 'form'. This form is characterised by three properties. First, the system must be consistent. Consistency means that in the system there are no truths that contradict other truths. There can be no paradoxes. Truths in formal logical systems are often called 'theorems'.

The second essential property of a formal logical system is called 'validity'; it means that the rules that we use in the system to produce conclusions cannot allow false inferences from true premises. The rules of the system are our guide to the truth, our logical compass. We are dependent on them. They are also like the cogs of a magnificent machine: they can always be depended on to perform the same operation, in exactly the same way, and producing – always – top-quality results.

The third property of a formal logical system requires that if a statement in that system is true, then it can be proven. This property is called 'completeness'. A formal logical system must be complete. It must be able to do everything by itself, including proving all its true statements. The converse must also happen: if you can prove a statement then that statement must be true.[6] Characterised by consistency, validity and completeness, a formal logical system resembles an intellectual machine, which, whenever you feed it with true facts, will spit out the truth and nothing but the truth. Formal logical systems are the foundations of mathematics, of logic and of science. Everything we know, or think we know, depends on them.

By defining formal systems mathematicians and logicians felt that they had achieved something great, perhaps the greatest intellectual feat that humankind had ever achieved. However, one man was still worried: the German mathematician David Hilbert (1862–1943), one of the most influential minds in modern science and mathematics. Hilbert was very concerned about the foundations of mathematics and logic and wanted to ensure they were based on solid ground. He was the kind of person who was never content about something until he had tested it to see if he could break it. In 1928 he set a challenge to fellow mathematicians around the world. He asked them to prove that formal logical systems are consistent, valid and complete. It was not enough to define the properties and build a mathematical theory. He asked that the mathematical theory about formal logical systems be proved to be true.

What Hilbert really wanted was for mathematics to be like science. In science, a theory is proved empirically by means of testing, experimenting, collecting data and verifying the theory's predictions. Hilbert wanted to test mathematical theories in a similar way. But the only way to do this was through mathematics. So, in effect, he was asking mathematicians to use mathematics in order to prove a mathematical theory. The close relationship between algebra and logic that began with Boole, and was made more intimate with Frege, had now reached a complete union. After Hilbert, 'metamathematics' would become the mathematical study of mathematics, logic examining itself – a very cybernetic idea indeed!

The machine that always tells the truth

I sometimes imagine Hilbert, as a child, reading *Gulliver's Travels* – and an idea from the book entering his subconscious and emerging the day he set his famous challenge. Swift's popular book, first published in 1726, was read by almost every educated European of the time. A particular scene in Book III might have caught the young Hilbert's imagination, a scene that quite possibly inspired events many years later that shattered logic and mathematics forever. Following his brief captivity by pirates, Gulliver is abandoned on the continent of Balnibarbi. After a visit to the flying island of Laputa, he is taken to the Academy of Projectors in Lagado, where 'useless projects' are undertaken. There, he is given a demonstration of a word machine, which is nothing less than a giant mechanical computer used for making sentences and books. Swift had fun with the satirical idea that a machine could render obsolete any study or expertise. Any idiot could write a masterpiece by cranking the Machine of Lagado. Young Hilbert may have been enthralled by the Swift's idea, and he may have also wondered whether such a machine could ever be built. The Machine of Lagado was capable of infinite logical combinations of words, similar to the brain of a human writer.[7] We could build one if only we could somehow encode (in a mechanical, electronic, or other way) every true logical statement about the English language, its grammar, syntax and use. If we could do that, then – should our world accidentally go to wrack and ruin – we would have a machine that we could simply crank up and it would reproduce everything that was lost. It would write every book, and reproduce all knowledge. A functioning Machine of Lagado would guarantee the eternal survival of our civilisation. It would be the formal system of all formal systems, the mother of all logic. In effect, the challenge that Hilbert posed in 1928 tested the reality of Swift's literary fantasy.

Hilbert's problem is called *Entscheidungsproblem*, German for 'decision problem'. It tests the completeness of a formal logical system. Simply put, it involves establishing a set of logical steps (called 'algorithm') that will accept a statement as input and then provide a 'yes' or 'no' answer depending on whether the statement received is true. If one could build a machine that executed that algorithm, then the machine could prove every theorem. It would be the perfectly logical machine, and would always tell the truth. It could also be the basis for a machine that could rewrite every book ever written, or could ever be written: the logico-mathematical Machine of Lagado.

Mathematicians around the world picked up the gauntlet and knuckled down to working out a solution. Amongst them was a young Fellow at King's College, Cambridge, called Alan Turing.[8] His solution to the problem would constitute an act of sheer brilliance that would ensure the young English mathematician global recognition. But when Hilbert set his challenge, Turing's moment of glory still lay several years in the future; we'll return to him shortly. First, let us a pay a visit to the barber.

The barber paradox

The reason Hilbert was so worried about the foundations of logic and science was that, by the time he posed the *Entscheidungsproblem*, certain cracks had begun to appear in the grand edifice of formal logical systems. It had been noticed that one could create logical statements that were not true, which meant that some formal logical systems were not consistent. One of the most notorious such statements was Bertrand Russell's (1872–1970) 'barber paradox':

Suppose there is a town with just one barber, who is male. In this town every man keeps himself clean shaven either by shaving himself or going to the barber. Using predicate logic, we can denote what happens to the men who want to shave in the town, in exactly the same way we did with the earlier sentence about all Greeks being philosophers or drinking coffee:

$$\forall\, x \in X,\ P(x) \vee Q(x).$$

This time the notation says, in plain English: 'For every man x who belongs to the male population X in town, the man x either shaves himself P(x) or the man x goes to the barber Q(x).[9]

That's all very well, but who shaves the barber? You see, the barber is also a man in that town. And yet, this logical statement that holds true 'for every man in that town' (hence the symbol $\forall$) is not true for him. The barber cannot go to the barber since he is the barber. But if he shaves himself then he is going to the barber, which according to our logical sentence cannot happen: he can either shave himself or go to the barber, but never both! In other words, the logical system is inconsistent. We have a logical sentence that is not true. Hilbert was horrified.

There are many other self-referencing clauses that cause similar logical paradoxes. For instance, a famous saying by the ancient philosopher Epimenides states: 'All Cretans are liars!' But if Epimenides, who was a Cretan himself, is a liar, then he is lying about what he is saying, and therefore is telling the truth about his lying, but since he is a liar we cannot believe him, and so on in a never-ending circle. Epimenides' paradox – as well as the barber paradox – occur because in language we have self-referencing clauses. Russell defined the problem of self-referencing paradox in mathematical logic like this: 'Does the set of all sets that do not contain themselves contain itself?' This was a terrible problem for the foundations of mathematics and logic, expressed more succinctly by Groucho Marx when he said: 'I do not want to belong to any club that will accept people like me as a member.' The club of mathematical truths was packed with lies. Someone had to get them out and clean the place up.

The two greatest British mathematicians and philosophers of that time, Bertrand Russell and Alfred North Whitehead (1861–1947), decided to solve the paradoxes, and ensure that the foundations of mathematics and logic remained sound and intact. Both strongly believed that if a statement were true, there had to exist a proof for it. Paradoxes, such as barbers, Cretans and Groucho's club membership, had to disappear if attacked with rigorous logical tools. Following Frege's example, they separated axioms from logic to keep logic pure. In their monumental quest for pure logic without contradictions, Russell and Whitehead used metamathematics to construct a theory that would allow all mathematics to derive from purely logical axioms.[10] The theory was meant to prove that formal logical systems were complete, i.e. that all true statements in the system could be proved to be true. Using Frege's notation system and the ideas of predicate logic, the two mathematicians laboured over writing one of the most important books in the history of mathematics and philosophy, a three-volume work entitled *Principia Mathematica*. The book was first published in 1910, and then revised and reprinted several times as the authors tried to make their arguments ever more watertight, the last edition coming out in 1927.

Principia Mathematica is perhaps the most famous book to have been read by only a handful of people, as Bertrand Russell often enjoyed quipping. Years after the book's publication, he suggested that perhaps only six people had ever read the whole thing! One of them was Wittgenstein, who tried hard to ensure that Russell and Whitehead succeeded, and offered many valuable comments and suggestions. Nevertheless, the book's intellectual impact has been stupendous. It established mathematical logic as a philosophical discipline. Using the notational and conceptual tools of *Principia Mathematica*, philosophers have henceforth been able to develop new ideas in epistemology and metaphysics. Computer science, linguistics and psychology borrowed much of their technical apparatus from Russell and Whitehead's pioneering work. In Artificial Intelligence, natural language is modelled using the advanced forms of predicate logic that these two brilliant mathematicians developed in their book. And yet – too bad for Russell and Whitehead that the *Principia Mathematica* did not solve the fundamental problem that Hilbert was so worried about. No matter how hard the two authors had tried to resolve paradoxes, doubts persisted about the completeness of formal logical systems. Paradoxes simply refused to disappear. Worse, a young mathematician from Austria came along in 1931 and proved that the problem of completeness was, in fact, unsolvable.

An Austrian comes along . . .

When Kurt Gödel (1906–1978) was a small boy in Brünn, his hometown in part of what was then the Austro-Hungarian Empire, people used to call him 'Mr Why'. He was curious about everything – languages, philosophy, metaphysics and religion – but it was mathematics and logic that would shape his destiny. At the age of eighteen, he went to the University of Vienna and joined the Vienna Circle through his supervisor Hans Hahn. Four years later, Hilbert posed his problem of completeness and, at the suggestion of Hahn, young Gödel chose the problem as the subject of his doctoral thesis.

The problem of completeness was Hilbert's *Entscheidungsproblem* stated in a more general way: are the axioms of a formal system sufficient to derive every statement that it is true? In other words, if one 'cleans' up a formal system of all paradoxes, chucks them all outside, and seals the doors with axioms, is the formal system now complete? By 1927 Russell and Whitehead, with their *Principia Mathematica*, had answered the question with a resounding 'yes' – or at least so they thought. Besides, this was the answer that every mathematician expected. The whole edifice of mathematics was based on the ancient belief, and conviction, that all true statements could be proved. Unfortunately, Gödel looked into *Principia Mathematica* and proved that they were all wrong. In his own landmark publication,[11] he proved that sufficiently complex formal systems couldn't be consistent and complete. His incompleteness theorem blew apart the foundations of mathematics and logic. What had once been solid ground was turned overnight into quicksand by an elegant proof that is still a monument to human ingenuity.

Gödel proved his theorem by using many of the metamathematical tools and ideas pioneered by Frege, as well as by Russell and Whitehead. In Part II of this book, I mentioned how Gödel used the reflexive method of logical substitution to prove his theorem. It is now time to look more deeply in how he actually did it. His genius idea was to code formal expressions as natural numbers, a logical substitution process that has since been called 'Gödel numbering'.[12] For example, he substituted v (the symbol for 'OR') with the number '2'; and ∃ (the symbol for 'exists') with the number '4'; and so on. He then used these expressions to construct an unprovable formula. In fact, what Gödel effectively did was to take Epimenides' liar paradox and formalise it.[13] This formalised statement was false if it was provable. By doing so, he contradicted the definition of a formal logical system: since the formula existed the formal system was incomplete. In a sense, he used the self-recursive paradoxes that Russell and Whitehead had tried so hard to hide under a mathematical carpet as the core argument of his proof. If a system contained paradoxes, and was therefore inconsistent, it was complete. If it was complete it could not be consistent. You could not have both. If you were a formal logical system[14] you could not have your cake and eat it.

It is impossible to understate the historical and cultural significance of Gödel's 'incompleteness theorem'. Our contemporary culture of relativism and tolerance for even the most absurd of notions of others, the way we understand and communicate our own values and notions, our collective reflex against the forceful imposition of our cultural values upon others whose ways we may find deeply abhorrent, are informed by Gödel's extraordinary achievement. Together with Heisenberg's uncertainty principle,[15] Gödel's incompleteness theorem is arguably one of the defining moments of the birth of postmodernism. These two historical conclusions of modern science and logic showed that there were unyielding limits to what we could know and, therefore, that there could never be absolute knowledge. Subjectivity was therefore confirmed forevermore. The intellectual revolution that began with Descartes in the eighteenth century was completed in the 1930s with quantum physics and Gödel's theorem. Heisenberg showed through the uncertainty principle that we could never know both the position and the speed of a fundamental particle; which meant that nature would remain elusive and intangible regardless of what we did or how smart our measurement instruments became. Gödel revealed that we could never prove the truth of every true statement; in effect, that there will be truths (or true statements) that we must accept by just believing. In the wake of these two shocking discoveries, no one could ever assert absolute authority over others, because such an authority was simply unattainable. Our universe prevented it, and so did logic. Henceforth, we would have to make peace with our intuitions and our ignorance.

But what of the truth machine suggested by Hilbert's *Entscheidungsproblem*? The one that always told whether a logical statement was true or not? Gödel's incompleteness theorem had shown that in a formal system there were true statements that could not be proved. Nevertheless, the *Entscheidungsproblem* also had to be proved. What if the machine had infinite time to work on a problem? Couldn't such a machine, by following logical, algorithmic steps, finally arrive at a proof? Many in mathematics and logic, including Hilbert, were still clinging on to the hope that an algorithm could beat Gödel's horror-inspiring incompleteness theorem. Their hopes were dashed forever in 1936 with the publication of Alan Turing's paper on computable numbers.[16] Gödel buried the omnipotence of logic, but it was Turing who placed the tombstone over its grave forever.

Turing's research was inspired by Gottfried Leibniz, the seventeenth-century German mathematician who had first dreamed of building a machine that could manipulate any symbol. He then took Gödel's theorem and reformulated it by replacing Gödel's arithmetic-based notation with simple, hypothetical devices that became known as 'Turing machines'. A Turing machine is like a tape recorder. The tape can move in both directions, back and forth, and the machine can record symbols on the tape by following simple instructions. There is also a 'state register', something we would today call 'memory', which keeps track of what the Turing machine has been up to. Remember Gödel's genius idea of substituting logical operations and expressions with numbers? In turn, Turing substituted logical operations and expressions with Turing machines. A Turing machine took a statement as its input, applied a logical expression (or 'formula') that was written as a set of instructions, and produced a binary result: the statement was either true or false. The first thing that Turing showed was that a Turing machine capable of performing a mathematical computation was equivalent to an algorithm (i.e. a series of logical steps that processed a statement and arrived at a conclusion). This is one of the most important observations in computer science. It means that all computers are in fact reducible to algorithms – I will return to the significance of this a little later.

But then Turing went on to show something even more earth-shattering: that it was not possible to determine algorithmically whether a Turing machine would ever halt. By 'halting', Turing meant that the machine would come to a conclusion and thus end its operation. So imagine a Turing machine beginning to work on a problem, say a logical statement. It starts applying a set of instructions to examine if the logical statement is true or not. And keeps going, and going . . . without end. The Turing machine is stuck in an eternal 'hmmm' moment. It cannot give an answer. It cannot tell whether the statement is true or not. It does not 'halt'. Turing proved that it was impossible to know in advance whether the machine would halt or keep on going. To rephrase it more mathematically, Turing showed that no 'formal language' (what we would today call a 'computer language') exists that can manipulate any series of symbols and determine the truth of mathematical statements.[17] The answer to Hilbert's *Entscheidungsproblem* was found to be a dismal one, for there was no solution to the problem.

Turing's proof came as the *coup de grâce* for all of those who hoped that logic was grounded on sound foundations. And he then went further. By inventing Turing machines he sparked the computer revolution that would follow. Computers are algorithms, and Turing showed how to build the mechanical 'algorithm of algorithms', or the 'Universal Machine'. Thanks to him, engineers could now proceed and built computational machines that could perform any logical processing and thus solve any problem. But Turing also posed a fundamental question about the future of computers: since they were limited by their mathematical nature, how could they ever be 'as intelligent' as human brains? Surely we humans are not reducible to an algorithm. Surely we are more than the sum of our parts. Or are we not?

Is Gödel the prophet of doom for AI?

Computers are programmed using symbolic programming languages that are based on logic. Indeed, computer languages are an alternative way to denote logical statements and logical relationships. In their syntax, modern computer languages use Frege's quantifiers, as well as all other tools that logic provides, in order to deduce new knowledge from existing facts, or use existing facts to execute some function. Turing showed that all computers can be reduced to an algorithm. The problem with this algorithm is that it is severely limited when compared with the human brain. It is bound by Gödel's incompleteness theorem. It does not possess 'intuition'. It cannot make leaps beyond the data it processes, as we do when we arrive at a 'eureka!' moment. Computers cannot cry 'eureka!' Where we humans would start jumping out of our baths and running naked in the streets of Syracuse, computers would get stuck in an endless loop and never halt. Or so it appears.

British physicist and mathematician Roger Penrose believes that Gödel's incompleteness theorem and Turing's subsequent solution to the *Entscheidungsproblem* spell the doom of Artificial Intelligence. In his 1989 book *The Emperor's New Mind*,[18] Penrose notes that, as Gödel proved, the algorithm cannot exist that can prove its own incompleteness. And yet Gödel, a human being, did exactly that. Gödel's intuition invented a numbering system and a proof for his theorem. Therefore, Gödel's brain is not running an algorithm. Since all modern computers can be reduced to an algorithm it follows – by deduction – that human brains are not computers. The converse is also true: algorithm-based computers cannot do all the things human brains are capable of. Something will always be amiss. There will never be true Artificial Intelligence. A computational theory of mind, the theoretical basis of traditional AI, is therefore false.

Penrose also noted that the reason why traditional computers fail to attain the capability for intuition that a human mind has, is because they are deterministic. A simple definition of determinism in this context is that computers will always produce the same output[19] for a given set of inputs. On your PC one plus one always make two. And yet, human brains can come up with various other outputs for that same set of inputs. For instance, we might argue that one and one never make two because neither of the two numbers added is 'two': they make two 'ones'.[20] We saw how Penrose and Stuart Hameroff elaborated their idea of non-deterministic brains, and arrived at the conclusion that our consciousness must be based on quantum physics. The debate continues around their theory, although I belong to the deeply sceptical side. Nevertheless, Penrose's argument about the limitation of determinism in classical computer architectures is a very valid one, and has to be dealt with. Let's analyse it a little more.

Turing was also aware of the limitations of Turing machines compared with human intelligence, and was not too happy either. In his 1938 PhD thesis at Princeton,[21] he introduced the concept of 'oracles' in order to deal, indirectly, with the issue of 'incomputable' intuition. Turing was trying to find a back-door exit from the confines of Gödel's incompleteness theorem. He mused that if there was a theorem that could not be proved in a formal logical system then this theorem could be called an axiom. An axiom – by the definition of the word – does not need a proof. This way one solved the problem of completeness by creating a new logical system with its own improvable theorem, and so on. However, if one kept doing that, this process of iteration led to infinity. In his thesis, Turing examined what it means to keep iterating logical systems to infinity, by continuously renaming their improvable theorems as axioms: one ended up with a super-formal system with an infinite set of axioms. Could that be a better computer model for how human consciousness works? To build this computer model in practice Turing suggested that classical, algorithmic machines should be augmented with 'oracles': these are machines that can decide what is undecidable by a normal Turing machine, for instance the halting problem. But what would an oracle machine look like? And how would it function? In theory, oracle machines are just like Turing machines, with the additional ability to answer 'yes' or 'no' whenever the normal Turing machine cannot find the answer. This, effectively, transforms an improvable theorem into an axiom. Nesting Turing machines with oracles produces a hypercomputer. You can go beyond that too: you can start nesting hypercomputers together, all the way to infinity. And that would be a replica of the human mind.

The provocative, and very interesting, thing about the concept of oracles is their randomness. The 'yes' or 'no' answer can be random. It does not necessarily require 'reasoning'. The machine oracle can be like a simple on/off switch. The randomness of the oracle takes care of the determinism of classical Turing machines. No given set of inputs will produce the same set of outputs. The oracle's answer 'yes' or 'no' could happen by chance alone. A different way to visualise oracles is as a decision tool used to navigate through a maze. When there is no logical way to infer which way to go, you hit the switch and let the switch decide which of the two ways in front of you to take. In a way, Turing's oracle machines resemble the decision-making mechanism used by the protagonist in *The Dice Man*, the 1971 novel by George Cockcroft, writing under the pseudonym Luke Rhinehart. In the novel the hero, a psychiatrist also named Luke Rhinehart, begins to make life choices by casting a dice. If Luke were an algorithmic-based computer the dice would be the oracle machine. Turing argued that this model was the closest to how actual, natural, consciousness works. Somewhere in our brain there is a dice that rolls. Often we choose to call that chance happening 'free will'.

Bailed out by paradox

There is also another way of viewing Turing's oracle machine concept. Classical Turing machines are simple calculators. In today's terms we can regard them as offline batch computing processes: a stand-alone computer performing a computational process on a data set. But once you start connecting this computer with an external database, or an external computing process, interesting things begin to happen. The offline computer comes online: it can interrogate an external database, or stop its process and let another machine take over. Online computing is a model of Turing's oracle machines connecting to classical Turing machines. The Internet, where billions of computers interconnect and query one another in a continuous, dynamic and non-deterministic fashion, is a realisation of Turing's infinite hypercomputer. Viewed from this angle, the ever-evolving and expanding Internet appears to transcend the problem of incomputability.

Let's return to Penrose's argument that consciousness could never be coded in an algorithm, i.e. in a computer. He is right to refute the ability of a stand-alone classical computer to achieve consciousness. However, when it comes to an interconnecting and ever-expanding network of computers his refutation begins to weaken. Indeed, as the Internet connects not only machines but also people who interact with the machines, the end result is a cybernetic hyperorganism with all the characteristics of a second-order cybernetic system. A fundamental property of such a system is self-referencing. In such a system logical self-referencing paradoxes have the capacity to provide meaning to meaningless processes – as we saw in the previous part of the book. Does that mean that the Internet is potentially sentient?

There are many who think so. But perhaps there is a way to settle the matter, and get an empirical answer to this question. If the Internet has any degree of sentience it ought to exhibit signatures of consciousness similar to those discovered by Dehaene and his team. We only need to find a way to 'brain scan' the Internet, and examine the images. Such images could be obtained using current maps of Internet interconnectivity and measure the amplification of information that flows through. For instance, let us go back to the dark day of Tuesday September 11, 2001, as the news of the terrorist attack on the Twin Towers of the World Trade Center began to spread around the world. As the wave of information propagated and amplified did the Internet become aware of what was going on? It would certainly be an interesting experiment to carry out.

We thus arrive at a very interesting result. Instead of Gödel's incompleteness theory dooming the ultimate goal of Artificial Intelligence, his deep insight on the importance of self-referencing bails Artificial Intelligence out of incomputability. As Hofstadter notes,[22] Godel's numbering system makes it possible for any formal system to spew truths about its own properties and therefore become, in a manner of speaking, 'self-aware'. Meaning emerges as the formal system folds into itself. Russell and Whitehead regarded paradoxes as weaknesses, but Gödel showed us that these paradoxes are what make formal systems capable of achieving meaning. Research results from studying the neuronal processes that cause consciousness in the brain seem to reflect Gödel's insight. If we abstract Dahaene's 'signatures of consciousness', and rephrase them in computational terms, our brain is a massive network of densely interconnected Turing machines. Each of these tiny machines (our individual neurons) processes a set of inputs (electrochemical excitations entering their axons) either by amplifying or reducing the strength of these excitations as outputs. This relatively simple mathematical process of signal integration, augmented by the massive computational parallelism in the brain, arrives at a stage when neuronal patterns that have spread across many anatomical areas of the brain mirror the brain's mirroring of the world. The brain thus enters into a self-referencing loop, and becomes self-aware. Thus, the computational model for Artificial Intelligence is not completely false. We do not need quantum physics to explain non-deterministic phenomena in the brain, or to construct machines that can achieve meaning out of meaningless symbols. We can use classical physics in the macroscopic world of digital electronics to engineer machines with self-awareness. Nevertheless, the computational model of Artificial Intelligence needs to be developed and expanded a lot further from where it sits today; and embrace, and exploit, the Gödelian power of self-referencing.

To see how this may happen, we need to go back and trace the history, and evolution, of computers. We need to examine that singular moment when the centuries-old human dream of building a thinking machine escaped from the pages of literary fiction and entered the engineer's workshop. Back to our time machine – destination Victorian London.

13
THE PROGRAM

On the second floor of the Science Museum in London the interested visitor will come across a section dedicated to the history of modern computing. The section begins with a display of a nineteenth-century computational machine that was designed and built by one of the most celebrated prodigies of Victorian England, the eminent mathematician, philosopher, inventor and mechanical engineer Charles Babbage (1791–1871). Arguably, this is a contentious proposition by the curators. If this were a section dedicated to evolutionary biology many scientists would be aghast. Presenting Babbage's Difference Engine as the starting point in the evolution of computing is akin to a creationist's account of human evolution, like showing Adam and Eve having sprung forth *ex nihilo* by the grace, and genius, of the superior being that created them. We know that the biblical version of humanity's origins is wrong. Like human evolution, modern computers, too, came about as the result of a complex, fragmented and totally haphazard historical process. Taking the biological metaphor further, the DNA of modern computing is a symbiotic relationship between two strands of completely disparate ideas,[1] fused together by the changing economic and social conditions of the past two centuries. The first idea stems from our primal and innate desire to create artefacts that behave as if they were alive. The second idea comes from the study of logic, and the curious discovery that logic and mathematics are twins. Let's look at these two ideas and how they became one.

We saw how automata, imported from Byzantium and the Caliphate, became popular in Western Europe during the Renaissance. By the late 1700s the French inventor Jacques de Vaucanson (1709–1782) had designed and built the first automaton purportedly capable of digestion, the 'digesting duck': one could feed the mechanical duck with kernels of grain and the machine seemed to metabolise and discharge them through defecation. Many other inventors and engineers fashioned entertaining automata. In the early 1800s the Swiss mechanic Henri Maillardet (1745–1830) created an automaton capable of drawing four pictures and writing out three poems. What automata did was imitate nature. In effect, they were the first artificial cybernetic systems. And although mechanical engineers would become increasingly good at creating ever more sophisticated automata by the late 1800s, their use remained limited to entertainment. During the 'golden age of automata', from 1860 to 1910, thousands of clockwork automata were exported from the workshops of Paris to many countries around the world. Most of them were mechanical singing birds.

Meanwhile, the Aristotelian concept of logic as a process begged the question as to whether such a process could be automated. Gottfried Leibniz toyed with this question several years before Boole. Like many learned Europeans, Leibniz was aware of the famed 'Doctor Illuminatus', the Catalonian philosopher and mystic of the Middle Ages Ramon Llull (1232–1315), who was the first person to conceptualise a logical machine. In his book *Ars Magna* of 1305, Llull describes a paper machine made of several interconnected and concentrically arranged circles. Having invented a symbolic alphabet to represent the qualities of the divine, Llull used the machine to describe preliminary truths which, when combined, could answer any possible question. His purpose was to use such a machine in order to convince Muslims that Christianity was a superior faith.

Leibniz was unimpressed by the missionary zeal of Llull, but very intrigued with the idea of logic producing new knowledge through a mechanical process. His boundless curiosity led him to the invention of calculus, the binary numerical system, and more. In 1679, he dreamed of a machine in which binary numbers were represented by marbles. These marbles moved down cards punctuated by holes. By manipulating the holes and the movement of the marbles he showed how mechanical computation could be automated. In essence Leibniz postulated that any computational machine ought to somehow have moving parts that could be manipulated, stored and counted. Modern computers are very similar to what Leibniz imagined. Pulses of electrons, like marbles, move around thanks to differences in electrical voltage; modern shift registers controlling inputs and outputs in order to shift data are representations of Leibniz's punched cards.

Around the same time in France, the mathematician Blaise Pascal (1623–1662), together with the German astronomer Wilhem Shickard (1592–1635), developed a mechanical calculator for practical applications. Pascal was a teenager at the time and was trying to help ease his father's workload, his father being a tax collector in Rouen. After much experimentation the 'Pascaline', as it became known, was presented to the public by Pascal himself in 1645. This was a mechanical calculator that could add and subtract any two numbers by means of five input wheels marked with numerals from 0 to 9. The Pascaline was based on five identical rotating mechanisms connected in a series. Each mechanism was made up of two cram wheels that translated an arithmetic input from the user into a visual output on top of the machine. Every time the user turned the input wheel beyond 9, a 'carry 1' mechanism triggered the next mechanism in the series to show the next digit in its respective dial. By today's standards it was cumbersome to use, but it made perfect sense when one needed to add quickly very large numbers in the mid-seventeenth century. Because the 'carrying mechanism' could only take place in one direction (i.e. the machine could only add numbers) subtraction was performed using a mathematical technique called 'the 9's complement'. The '9 complement' of any number 'a' is '9 minus a'. By cleverly translating the numbers to be subtracted into their '9 complements' the Pascaline added the latter, and thus performed the subtraction between the original numbers.

Leibniz added multiplication and division to Pascal's machine. In 1685, he designed the pinwheel calculator and the 'Leibniz wheel', which, when placed in a certain arrangement, could achieve these processes, too. This was made possible thanks to the way the pinwheel was designed, with teeth of incremental lengths that meshed with a counting wheel. Leibniz's wheel, as well as the wheels-in-a-series mechanism of the Pascaline, would later become the key mechanical elements in the 'arithmometer', the first mass-produced mechanical calculator that was finally built in 1851. The device sold millions and was widely used in offices around the world. In fact, Leibniz's inventions were used to automate calculations until the advent of electronic calculators in the mid-1970s.

Several other engineers and mathematicians experimented with mechanical calculators and machines during the eighteenth and nineteenth centuries, including Lord Stanhope (1753–1816) in Britain, who designed a pinwheel machine in 1775.[2] These machines were not automata per se but needed a human operator to turn the wheels and note the result. Nevertheless, the technologies and key mechanical elements that had been invented by the constructors of automata during previous centuries were now repurposed in mechanical calculators. This repurposing presaged the advent of robots in the late twentieth century, which would become the next stage in the evolution of human-like, and animal-like, automata. It also infused the culture of computing with anthropomorphism and inseminated it with the implicit expectation of computational machines having 'brain-like' qualities and behaving like humans; notions that will later resurface in Artificial Intelligence.

Babbage's Difference Engine, alongside the calculating machines of Pascal, Leibniz and others, was the precursor of what followed: automatic machines would meet, and fuse, with symbolic logic. This was a significant merging of technologies and traditions, a true 'singularity point' in the history of technology. Setting off from ancient Alexandria automata transformed into calculating machines by the late eighteenth century. Artefacts that were once used to simulate and mimic life became machines that crunched numbers. These computing machines were vital in order to satisfy the need for expedient and error-free arithmetic calculations in a world that changed rapidly and began to industrialise. At that particular historical confluence, symbolic logic had adequately advanced thanks to Boole and Frege. Symbolic logic met with this new breed of calculating automata and, with time, a new technology emerged: digital, electronic, general-purpose computers. The historical merging of those two different technologies, the mechanical-physical and the logico-mathematical symbolic, would determine the dual nature of modern computing, and its separation into hardware and software. Nevertheless, the evolution towards this separation was by no means a foregone conclusion in the late eighteenth century. As it turned out, general-purpose computers had to be discovered twice.

The wheels of industry

Charles Babbage was the first person to produce an engineering design that distinguished between a program and the machine capable of executing it. For this he is rightly revered as the father of the modern computer. He did not arrive at this innovative design all at once. He spent years working on a number of different things, until he had the brilliant idea of borrowing a technological solution from the textile industry and applying it to automating solutions of general problems in mathematics.

Babbage was born in London in 1791 at a time when the Hanoverian kings of Great Britain had laid the foundations for one of the most liberal democracies in Europe. Democracy, property rights and the rule of law became engines of unprecedented wealth in the new world order that was taking shape in the wake of the Industrial Revolution. By the early nineteenth century, sea lanes, railways, international banking and trade interconnected every habitable continent on our planet, for the first time in history. According to analysis by anthropologist Ian Morris,[3] the Industrial Revolution is the most significant event in human history. Everything about our world changed after that, including the level of sophistication in our social organisation, the ways in which we build our cities and fight our wars, how we share and process information, as well as the amount of energy we can harness and use.[4] The key technology that ignited the Industrial Revolution was the steam engine. James Watt's ingenious improvement of the efficiency of the steam engine, introduced in 1775, transformed the global economy by mechanising labour. Many historians refer to this period as the 'first machine age', when machines became an integral part of human society and changed it forever. By the time Charles Babbage came of age, Great Britain, the first country to industrialise, was the unchallenged imperial, economic and naval power. The coronation of young Queen Victoria in 1837 sealed Britain's supremacy and ushered in an era of unprecedented wealth and scientific endeavour. At the heart of Britain's prosperity was manufacturing, and in particular the manufacturing of textiles.

Babbage was no stranger to manufacturing. In 1832, he published a book[5] about the organisation of industrial production that made him famous. In the book, he explored how best to organise industrial production, and discussed forward looking ideas such as division of labour and profit sharing, as well as the rational design of factories. It was the result of many years of visiting factories and studying manufacturing processes. It was also this knowledge and insight that inspired Babbage to design his first machine that mechanised arithmetic: Difference Engine No. 1.

There are striking commonalities between arithmetic calculations and manufacturing processes. Both can be broken down into small, self-contained units that can be processed separately, or in batches. This 'batching' simplifies the complexity of the overall process and makes it manageable. It also means that the outputs of one unit are the inputs of the next. A manufacturing process is very much like a mathematical, or logical, sequence of steps, i.e. an algorithm. Moreover, manufacturing processes are recursive: they repeat again and again, so that multiple copies of a product get produced, en masse. Quality is also vital in a manufacturing process, and is best preserved by checking that each self-contained unit of processing delivers the best possible output – before it becomes the input of the next step.

Based on such observations, Babbage concluded that numbers could be manipulated in a manufacturing-like process that produced new numbers through an algorithm. Each calculation would be a step in the algorithm and provide a partial result as an output that could be checked for errors. This way, arithmetic errors – a reflection of the 'quality' of computational processing – could be minimised. Babbage also noticed that the use of steam power that automated machinery in factories could likewise be used to automate computational machines. His vision for the future included automatic machines that computed numbers and never failed to produce the correct answer. As in the manufacturing of goods, the manufacturing of calculations would be mechanical, powered by steam engines, free of human error and capable of producing infinite copies of the same perfect thing, always.

It was the Astronomical Society in London that provided Babbage with the impetus for putting together a computational machine inspired by manufacturing. The

Society was commissioned by the British government to check whether the Nautical Almanac, used by navigators and sea captains, was reliable. Naval navigation – essential to trade and foreign policy – depended on the accurate calculation of longitude. In turn, this was based on tabulations of lunar distances calculated by human 'calculators' who used logarithmic rulers. The Society discovered that these logarithmic and trigonometric calculations were error prone; and that too many discrepancies could occur whenever different human calculators worked out the same tables. There was a need to 'objectify' and standardise these calculations, as well as making them as accurate as possible. Babbage, who was a member of the Astronomical Society, and knew about the mechanical calculators of Pascal and Leibniz, set out to design and build a British calculating machine.

At that time, the key problem with machines designed to calculate logarithmic and trigonometric functions lay in division and multiplication, neither of which could be carried out by the 'Pascaline' or its clones. In order to overcome the problem Babbage used the properties of difference equations. These equations define a recurrent relation between two variables. By using the output of a calculation as the input of the next calculation, a difference equation can perform the equivalent of division and multiplication by simply adding or subtracting two numbers. The use of difference equations as the way of constructing the Difference Engine was of great significance. It became apparent from the start that recursive functions, as well as the concept of a repeated operation – i.e. a 'loop' – were inextricably linked to mechanical computing.[6] The fundamental mathematical expression of cybernetic behaviour and complexity would thus become inextricable from the evolution of computing, marking a complete departure from the simple, linear computational machines of Pascal and Leibniz.

Babbage was purportedly a rather difficult person to work with, and he ended up falling out with his engineer Joseph

Clement over the costs of creating the Difference Engine in 1831. Only a prototype was built, which is now on display in the Science Museum, London. Nevertheless, he persisted and improved the original designs, which he called Difference Engine No. 2. It was this second engine that was finally built over a hundred years later, in 1991, and which performed, successfully, the operations[7] envisaged by its original creator. But Babbage's place in the pantheon of computing is not due to the Difference Engine. By 1837 he had already begun the design of a revolutionary new machine that could perform any calculation. Like the Difference Engine, this new 'Analytical Engine' – as he called it – was also inspired by manufacturing.

The Analytical Engine Singularity

During his studies of manufacturing in England, Babbage had noticed the significant innovations in the textile industry. In 1801, the Jacquard loom[8] used punched cards in order to automate the mechanical weaving of very complex patterns. In the separation of the mechanical part of the machine from the patterns to be woven, the Jacquard loom foretold the computing dichotomy between hardware and software; i.e. between the physical medium that executes the processing and the informational 'pattern' that is either the set of instructions for the processing, or the object of processing, or both. Furthermore, punched cards could be stitched together to form long tapes of patterns that were input and 'executed' by the looms. Babbage introduced the use of punched cards and tapes into computing for the first time. Punched cards would be reinvented a century later by computer pioneers, and tapes carrying symbols were to be used by Alan Turing in order to define the mathematical conceptualisation of the modern computer.

There were three kinds of punch cards in the Analytical Engine designs and these were read by three different readers: one for all four arithmetic operations, one for numerical constants and one for load and store operations. The Engine's output was a printer, a curve plotter and a bell. It had a memory that could store 1,000 numbers of 40 decimal digits each, which amounted to approximately 16.7 kilobytes. Interestingly, Babbage designed the Analytical Engine as a Difference Engine curved back upon itself. A good way to visualise this is by imagining two Difference Engines mirroring each other, the outputs of one becoming the inputs of the other. This reflexive design made it possible to add conditional logical branching statements such as 'if A, then B' as well as loops. Calculations could be repeated and the results branched off for further processing by the interlinked difference engines. It was a very cybernetic design indeed! These innovative, self-referential features made the Analytical Engine 'Turing complete', meaning that it was a machine that could simulate every other machine.[9]

Unfortunately, Babbage never completed building the machine of his dreams, although he tinkered with the designs until his death in 1871. Following his repeated procrastinations, the government funding he received was cancelled after the Treasury lost faith in him delivering anything at all. Nor did he write a manual to describe how the Analytical Engine functioned. However, in 1842 the Italian mathematician and military engineer Luigi Menabrea (1809–1896) wrote a description of the machine in French. One year later Countess Ada Lovelace, the daughter of Lord Byron, translated the description into English. Included in her translation was the first computer program ever written, which makes Ada Lovelace the inventor of software.

Ada Lovelace (1815–1852) was the only legitimate child of Lord Byron. A month after she was born, her father abandoned her and her mother, and travelled to Italy, where he would ultimately meet up with Percy Bysshe and Mary Shelley by Lake Geneva the following summer. Ada never saw her father again. Lord Byron would die in Greece eight years later, while taking part in the Greek War of Independence.

Although a sickly child deprived of parental affection, Ada proved to be very talented at mathematics and possessed of a keen curiosity for technology. At the age of twelve she developed an interest in flight, and began to experiment with artificial wings, studying the anatomy of birds and writing a book entitled *Flyology* in which she illustrated her ideas and findings. She first met Charles Babbage in 1833, and became fascinated with the Difference Engine. Babbage was also impressed by her mathematical acumen and used to call her the 'Enchantress of Numbers'. By annotating her translation of Menabrea's description of the Analytical Engine,[10] Ada Lovelace gave us one of the most significant documents in the history of computing. Impressed with the potential of her friend's design, she wrote: 'Mr Babbage believes he can, by his engine, form the product of two numbers, each containing twenty figures, in three minutes.' She then proceeded to demonstrate how the Analytical Engine would work by writing an executable algorithm that could calculate the sequence of the Bernoulli numbers.[11] This algorithm is considered to be the first software program ever written.

The Analytical Engine and Ada Lovelace's 'first program' have been subjects of fascination for historians and fiction writers alike. How would the world be today if the British Treasury had not stopped funding Babbage's dreams and designs? Would the Analytical Engine have ushered in the age of computing one hundred years earlier? These are, of course, questions for writers of 'alternative history', and indeed they have been explored in several novels, short stories and comics. Alas, in the universe we inhabit the Analytical Engine, despite its significance, remained mostly unknown. The main features of a modern computer's architecture were rediscovered nearly a century later. And so was the separation between hardware and software. In this sense, the Analytical Engine was a technological singularity that happened in a world not ready yet to make something useful of it. Similarly to Hellenistic innovations such as the Hero's Steam Engine and Hipparchus'

Antikythera Mechanism, Babbage's great invention was well before its time.

Nevertheless, Babbage's achievement is profoundly remarkable. He had invented a machine that could perform multiple functions without the need to reconfigure its mechanical parts. By separating hardware from software, Babbage created unlimited possibilities for computation. Arguably, the full significance and repercussions of this separation are yet to be fully comprehended, and in all probability it will take several more decades to do so. Like Watt's steam engine, which heralded the start of the 'first machine age' of the Industrial Revolution in the late eighteenth century, modern computers are currently ushering in the 'second machine age'[12] through the digital transformation of our economy and societies. They do so because they are general-purpose machines that can run any program. Babbage's intellectual leap from the Difference Engine to the Analytical Engine was the moment when the seeds of the second machine age were conceived.

It was also a moment that the ghost of Descartes must have delighted in, for dualism had found its way into computing. As algorithms became programs, mathematics was also transformed. In effect, a mathematical function (in Ada's case the algorithm that calculated the Bernoulli numbers) became 'alive': it did not simply describe the relationship between variables and constants, but now *did* something, too. Programs set physical machines in motion, guided their operation, and acted like a 'soul' would in an otherwise purely mechanical body. This historical transformation of algorithms and mathematical functions from static to dynamic was inherited by software, and to this date underlies the dualistic nature of modern computing. The ancient dichotomy between matter and form re-emerged. Matter is hardware. Form is software. In Ada's time 'software' retained its physical instantiation: it took the form of the punch cards that the Analytical Engine would have to read in order to execute the calculation. This 'physical instantiation' of the computer program would persist with the continued use of punched cards in early twentieth-century computers. But as computer engineering evolved, and punched cards were phased out, it became apparent that a program is just a 'pattern of information'. Indeed, as symbolic logic became computer languages, the 'program', as well as the 'data', dematerialised completely. Atoms, the units of matter, became bits, the units of information. The program and the data transmuted from mechanical clogs and cards into pure, immaterial form. And thus software – like a Cartesian *res cogitans* – took precedence over hardware, and became the ghost in the machine.

Computing reinvented

We saw in the first chapter of this part of the book how, ten years after Babbage's death, George Boole demonstrated the way in which thinking could be automated by means of symbolic logic. His discovery, and its subsequent expansion by Frege's predicate logic, laid the foundations of modern computer languages. However, it was Alan Turing who linked logic and computational machines forevermore: the 'Turing machine' is in effect an Analytical Engine that processes a strip of tape with logical symbols written on it. Processing is executed according to a table of rules – the 'program'.[13] The only difference between the Turing machine and what Babbage and Lovelace had in mind is that the Turing machine is comparatively indiscriminate about the symbols it processes. The symbols can include numbers, theorems, or any other logical construct. That's computer software!

Computer hardware, on the other hand, had a more nuanced upbringing, puberty and adulthood. Independently of Babbage, the American inventor Herman Hollerith took up punched cards as an idea for data storage in the late 1880s. He invented a number of other innovations, such as a processor for punched cards (the 'tabulator') and a key-punch machine.

His company undertook the 1890 United States Census with great success, and in 1911 became one of the three companies that merged to create 'International Business Machines' – or IBM for short. By 1920, electro-mechanical tabulating machines had mostly supplanted purely mechanical calculators. Punched cards had also become ubiquitous in industry and government, and were used for accounting as well as administration. In 1935, IBM punched-card systems processed the records of twenty-six million workers as the United States instituted its social security system.

From the late nineteenth century till the late 1930s a number of other, seemingly irrelevant inventions were made that, with time, would define modern computing. Two of them were, arguably, the most significant. In 1876 Graham Bell discovered the telephone. And four years later, in 1879, Thomas Edison discovered the incandescent lamp.[14] These two inventions would become fundamental in the birth and evolution of electronics and telecommunications. By the time Claude Shannon began to formulate his information theory, there was a pressing need for better telecommunications systems because of the ubiquity of the telephone. Meanwhile, Edison's incandescent lamp had evolved into sophisticated vacuum tubes that acted as electronic amplifiers, rectifiers, switches and oscillators. Shannon was the genius who combined logic with electronics. In 1937, as a twenty-one-year-old student at the Massachusetts Institute of Technology (MIT), he demonstrated how Boole's logic could be applied in electronic circuits to construct and resolve logical or numerical relationships. In effect, this meant that there was a direct correlation between Boolean logic and certain electrical circuits (now called 'logic gates'). Shannon's findings represented a brilliant leap in the history of technology. His Master's thesis was the foundation of the digital computer and digital circuit design. Every piece of the puzzle was now falling into place. The mathematical description of a general computation machine was given a year earlier by Alan Turing in his 1936 paper 'On Computational Numbers'. Thanks to Shannon and Turing, logic, mathematics, electronics and computers were coming together as one.

In the twentieth century, the steam of Babbage's time was replaced with electricity. This meant that it was now possible to power specific devices separately, which was a great improvement from having to connect every element of a mechanical machine to a single power generator. Nevertheless, it took some time for electricity to phase out the mechanical transmission of power. In the beginning the two worked together. The first electro-mechanical computers to be constructed used electric switches to drive the mechanical relays that performed calculations. In 1939, the German computer pioneer Konrad Zuse (1910–1995) created Z2, an electromechanical relay computer.[15] Around the same time, in the early decades of the twentieth century, analogue computers used mechanics and pneumatics to model natural processes: they were physical, rather than digital, simulations of natural phenomena.

Both these technologies, electro-mechanical and analogue, would be superseded in the very near future by faster, all-electrical, digital computers that used vacuum tubes and sophisticated electronic designs. The socio-economic conditions that were absent when Babbage designed his Analytical Engine were now present. As the 1930s drew to a close all the fundamental elements of the forthcoming computer revolution had been invented and were waiting for the opportune moment to fuse together in one big bang. What was missing was a singular event, a spark, and for affluent societies to begin investing heavily both financially and intellectually in these formidable number-crunching machines.

That spark ignited on 1 September 1939, when Hitler's Wehrmacht crossed the eastern borders of Germany and invaded Poland.

14
FROM BLETCHLEY PARK TO GOOGLE CAMPUS

In the years since the outbreak of war in Europe in 1939, the world has changed in unimaginable ways. Countries' borders have been redrawn. Totalitarian regimes have all but disappeared. Even China has adopted a mixed economic model, and only North Korea lingers as a reminder of the tragicomic extravagances of communism. Science and medicine have progressed in leaps and bounds. Average life expectancy has increased and millions have been raised from poverty thanks to globalisation and free trade. Within a span of three generations, we have arrived at the information age. My parents were born in a world without computers, whereas I programmed my first computer when I was eighteen. My son requisitioned my iPad when he was one, and started playing with it as if it was second nature. Computers and telecommunication networks are so intricately entwined with our modern way of life that our very existence has evolved: we have become post-human – creatures with dual identities, a physical and a digital one. In the second decade of the twenty-first century we live both in a three-dimensional world of atoms, as well as in a digital continuum of bits. The global trend is clearly towards further, deeper and a more accelerated enmeshment of our lives, economies and societies with the digital world.

This enmeshment, often referred to as 'digital transformation', is fuelled by an unprecedented degree of investment and rapid innovation. Hundreds of thousands, if not millions, of young entrepreneurs around the world use computers and computer technologies to disrupt business models that held sway for centuries. Think of Airbnb, which has placed digital dynamite at the foundations of the hotel industry; or Uber, which has done the same for taxis. Both companies have reinvented how their respective industries are making money by pulling down barriers to entry and allowing anyone who can drive a car (Uber), or has a place to rent (Airbnb), to become taxi drivers and hoteliers. The list goes on and on. Twenty-something start-up founders become billionaires overnight, demonstrating that in the twenty-first century everyone can become a Thomas Edison. The tremendous impact of information technology on the economy is well documented. According to research by the World Economic Forum,[1] digitisation has boosted world economic output by US$193 billion during 2012 and 2013, and created six million jobs during that period. Using a Digitisation Index that ranks countries on a scale from zero to one hundred, the consultants Booz & Company found that an increase of 10 per cent in a country's digitisation score fuels a 0.75 per cent growth in its GDP per capita. That same 10 per cent boost in digitisation leads to a 1.02 per cent drop in a state's unemployment rate. Governments and private investors are elbowing for a place on the bandwagon of the 'second machine age'. Meanwhile corporate behemoths such as Apple, Google, Amazon and Facebook yield extraordinary economic power. Some would say their power goes beyond the economic: with unhindered access to our personal data, including information about our tastes, habits, vices, consumer spend and friends, these companies can potentially control not only what we buy, but also what we do and how we think. Never before have so few had so much power over so many.

In this chapter, I want to explore the drivers of this societal, as well as anthropological, evolution. Although computing machines began – as their name suggests – as contraptions that automated arithmetic operations, they quickly became applied to just about everything. What are the unique characteristics of computers that make them so flexible, adaptable and intrusive? How did the transformation from the physical to the digital come about? Where does it lead us? And, finally, in the age of big data, search engines, social media, mobile apps and the Internet of things, what role is there for Artificial Intelligence?

'War is the father and king of all'

Daring to complement Heraclitus' famous quote,[2] I would add that ballistics and encryption were the mothers and queens of all computers. The world war of 1939–1945 was fought with aircraft that often had to be shot down from the ground or from a moving ship at sea, and with encoded signals that coordinated sophisticated military movements of naval, land and aerial forces. Hitting moving targets successfully, as well as stealing enemy secrets, required massive arithmetic calculations. Thus, the Second World War became a conflict fought not only by generals and soldiers, but by mathematicians as well.

In Britain, the Government Code and Cypher School (GC&CS) set up its main site fifty miles north of London, at Bletchley Park in the Buckinghamshire countryside. Its goal was to listen in to German military communications. An assortment of linguists, crossword puzzle experts, papyrologists, chess champions and mathematicians from Cambridge and Oxford joined its ranks.[3] Amongst them was the young Alan Turing. He would be instrumental in devising a machine that broke the Enigma code used by the German air force and navy. The problem that Turing and his colleagues faced was enormous: the Germans had built complex ciphering machines that changed their settings daily. It was thus almost impossible to decipher them. Nevertheless, a combination of good luck and German sloppiness aided the British cause. Polish mathematicians produced an Enigma clone that was made available to the British scientists. On the basis of this, Turing[4] designed an electromechanical device

– called 'bombe' because of the terrible noise it produced – that could predict some of the daily settings of the Enigma machines by replicating the actions of several Enigma machines wired together.

This idea of a machine (the bombe) effectively simulating other machines (the Enigma) is central to computer theory. When a machine can simulate every other machine this is called 'Turing complete'. We saw how Babbage's Analytical Engine was the first 'Turing complete' machine in the world. Our modern computers are also Turing complete. But to get from the Analytical Engine to modern computers required a giant leap in the early 1940s, as the British and the Americans fought the Germans in the Atlantic. Turing's bombe cannot be considered a precursor of modern computer architecture. However, there was another machine designed and built at Bletchley Park that could.

Tommy Flowers (1905–1998) was a friend of Turing and a colleague of his at Bletchley Park. He and his team were assigned the task of breaking another German cipher called 'Lorenz', used by the German High Command to send messages to its field commanders. Flowers designed and built 'Colossus', the world's first programmable electronic computer. Instead of being electromechanical it used vacuum tubes, and incorporated a logic unit that performed Boolean operations. Using complex electronics rather than the well-tested electromechanical relays was very risky, but Flowers managed to convince GC&CS of the merits of new electronic technology, got backing for his project and delivered the first machine – called Mark 1 – in 1943 using 1,500 valves.[5] A Mark 2 redesign that used 2,400 valves went into service at Bletchley Park on 1 June 1944, and immediately produced vital information relating to the imminent D-Day landings. Computers and military operations would go hand in hand from then on.

Meanwhile, on the other side of the Atlantic, the Americans were also developing their own electronic computational machines, culminating in the design and construction of ENIAC (1946),[6] an electronic, programmable and Turing-complete machine that was used for artillery-timing tables for the US Army's Ballistic Research Laboratory. ENIAC represented a watershed in computer history. Its architecture would form the foundation of all modern computing. The key figure in generalising ENIAC's architecture was John von Neumann, who at the time was involved in the Manhattan Project at the Los Alamos National Laboratory in New Mexico. Von Neumann was fascinated by the design of ENIAC, and wondered how the computer might be easily reprogrammed to perform a different set of operations – not involving artillery ballistics this time, but to predict the results of a hydrogen bomb explosion. Invited by the team that developed ENIAC to advise them, von Neumann produced a landmark report,[7] which described a machine that could store both data and programs.[8] The 'von Neumann architecture' – as it has hitherto been known – demonstrated how computers could be reprogrammed easily. Until then computers had fixed programs, and had to be physically rewired in order to be reprogrammed. Von Neumann's architecture allowed code in a computer to be self-modified. One could thus write programs that write programs, an idea that makes possible the host of automated tools that computer engineers have nowadays at their disposal, such as assemblers and compilers. And that's how the modern computer was born: a machine divided into hardware and software; a machine that could be programmed easily, and was capable of performing not only mathematical but logical operations as well; a machine that could simulate any other machine.

By the time the Second World War gave way to the Cold War, the Western Allies had already developed advanced computer technology. They would use it to computerise their arsenal and tactics, and ultimately to win the Cold War. When the Berlin Wall fell in November 1989, the West had such an advantage over the Soviet Union in computing power that it was capable of shooting down most Soviet intercontinental ballistic missiles in mid-flight. President Reagan's 'Star Wars' initiative[9] used complex operational research algorithms run on fast supercomputers to guide counter-attack weapons so effectively that the Soviet Union was rendered a military lame duck.

The investment and interest in developing computing technologies in the West was not driven by the profound military applications of computing alone. The industry quickly realised the huge potential of this new technology in increasing productivity and automating business processes. The 1960s saw the development of business-specific computer languages (such as COBOL)[10] that were applied in solving problems in finance and manufacturing. Hardware also developed in leaps and bounds, and by the early 1980s most businesses in the developed world used computers in various degrees. As electronics moved from vacuum tubes to solid-state transistors they started to become miniaturised. Soon computers that had once sprawled across several big rooms could fit in the palm of a hand. The invention of the microprocessor, a single chip that included all the circuitry of cabinet-sized computers, led to the proliferation of personal 'microcomputers' after 1975. And thus computers switched from being primarily industrial devices to become mass-market products providing a multitude of users with a plethora of applications. Scientists used them to test ideas; students to do homework; children and adults to play video games; big and small businesses to run processes, automate machinery and produce management reports. In 1982, *Time* magazine named 'The Computer' Machine of the Year.

And yet, by the early 1980s, the world had not seen the true beginnings of the social and economic transformation that computers would ultimately bring. Although businesses and universities had started to connect computers together in local networks that maximised processing efficiency and minimised storage costs, cyberspace resembled a vast ocean filled with millions of disconnected islands. The powerful drivers behind the computer's success were logic that provided automated solutions to most problems, and the separation between hardware and software that fuelled an incredible amount of innovation in both areas. Nevertheless, these two drivers could only take the world so far. The rhythms of life had remained more or less unchanged since the Industrial Revolution. Once you switched your computer off and left work you were disconnected – and free to return to 'normal' life. A clear separation existed between the digital and the physical. Computers were machines, in the same way that looms were machines. They performed specific work confined to their specific spaces, whether it was the office, the factory floor, or the playroom. Atoms still mattered more than bits. What was missing from making computers the facilitators of the information age was a third ingredient, one that would turn the disparate islands of computing into one huge, and ever-expanding, continent. Heraclitus' rumination on war was confirmed once again, after the cold kind of war between the US and the USSR begot the third missing ingredient of the information revolution, by fathering the Internet. But let's go back a few steps before that historical moment in the 1970s, and see the origins of Internet, and how the vision of a worldwide network that connected everyone with everything and with knowledge came to be.

La Mondothèque

The discovery of telecommunications in the mid-nineteenth century ignited the imagination of many. Telegraph wires began to span the world and connect faraway countries and regions together at the speed of an electric signal travelling down a wire. In *Paris in the Twenty-first Century*,[11] a novel written in 1863, Jules Verne describes a world in 1960 in which an international communications network spans the globe: one would write a letter and the network would transmit it and reproduce it exactly at the other end.[12] Verne had presaged the facsimile machine by extrapolating on the inventions of his day. By the late 1800s, pioneers such as Alexander Popov and Guglielmo Marconi had experimented with wireless telecommunications. These experiments caused the very concept of 'information' to change. Until the discovery of wired and wireless telecommunications, information had been stored mostly on paper and could be transmitted only physically: a book or dossier had to be carried from one place to the other. Inventions such as the phonograph, the photograph and cinema provided new media, other than paper, for storing information. Telecommunications demonstrated that information could now be carried electronically, and indeed that multiple copies of the information could be reproduced at the other end. What if one could combine all the information that existed – and was being produced – in the world with telecommunications? Wouldn't that create a library of all knowledge, available to everyone?

This was exactly the vision of a Belgian librarian called Paul Otlet (1868–1944). In 1910, Otlet envisaged the 'Mundaneum', the Library of Alexandria of the twentieth century.[13] It would include everything – newspapers, books, pamphlets, photographs, even audio recordings. Otlet devised an indexing system that could catalogue and aid search and retrieval of this information. The system would be realised in a Universal Bibliography made up of fifteen million index cards stored in filing cabinets. To reduce the enormous size of his Bibliography, Otlet advocated the miniaturisation of documents on microfilm, and designed automated search systems to locate information – a precursor of contemporary search engines.[14] All this information would be broadcast to users by radio, and stored in *la Mondothèque*, a workstation equipped with microfilm reader, telephone, television and record player.

Otlet was a typical political visionary of the early twentieth century and his vision of a centralised, highly managed and hierarchical structure for knowledge, and dissemination of knowledge, reflects the social climate of his age. It was that climate – inspired by Platonic political philosophy and mysticism – which ultimately led to the socialist, but also totalitarian political worldviews in the ensuing decades. The philosopher-kings of the modern, centralised *polis* were now incarnated in the servants and government of the state, which became responsible for delivering an equitable society, by force if necessary. Nevertheless, Otlet's efforts to realise his grand vision received lukewarm support from the Belgian government. Initially they offered him the Palais Mondial, a building in central Brussels, then later confined him to a corner of the building, only to finally evict him in 1924. In a sad outcome for a great vision, Otlet's collection was mostly destroyed by the Nazis in 1940. But Otlet never lost faith in his project. It is said that in his old age he could be seen piling up jellyfish on the beach, and then placing on top an index card bearing the number 59:33: the code for *Coelenterata* in his Universal Decimal Classification.

Paul Otlet is considered one of the visionaries of the information age, and yet his grand idea suffered from an obvious problem: what would happen to the Mundaneum if a great big bomb exploded on it? Wouldn't that spell the catastrophic end of all human knowledge? The destruction of Otlet's collection by the Nazis amply demonstrated the weakness of storing information centrally – exactly the kind of problem that the Defense Advanced Research Projects Agency (DARPA) of the United States Department of Defense was trying to solve in the late 1950s. Following the successful launch of the Soviet Sputnik and the realisation that the Soviets had cracked the construction of the hydrogen bomb, President Dwight D. Eisenhower and others realised that a nuclear war could potentially be fought using intercontinental missiles that could strike anywhere in the US. Such a strike might mean the destruction of the communications and control systems that the US military had in place to defend the country. A way therefore had to be found to make telecommunications indestructible in the instance of nuclear war. Telecommunication networks had to become decentralised and distributed, and guided by switching systems able to reroute traffic along whichever connections provided the optimal routes. In around 1965, DARPA commissioned the study of decentralised switching systems, which led to the development of the ARPANET[15] packet switching research network, which later grew into the public Internet. ARPANET sent its first email in 1971. Email was thus the Internet's first 'killer app'.

By the early 1990s, modems made email widely available. Computers began increasingly to connect to the Internet. The ocean was transforming into a new continent where information became a commodity. The invention of the World Wide Web ('Web' for short) by English computer scientist Sir Tim Berners-Lee provided a way for computers to share information. By Christmas 1990, Berners-Lee had built all the tools necessary for a working Web: the first web browser, the first web server and the first web pages.[16] The browser is one of the Internet's most widely used applications. It's what allows us to search and navigate through vast amounts of information. At the heart of the World Wide Web's information management system is the idea of hypertext. 'Hypertext' documents can be linked together through common reference words called 'hyperlinks'. A programmer can then use a computer language such as HTML[17] in order to code such documents, so that other programs can find them and read them. Perhaps the best way to think of this is through a short story written by the Argentinian novelist Jorge Luis Borges.

In 'The Garden of Forking Paths'[18] Borges describes a Chinese professor named Doctor Yu Tsun who spies on Britain during the First World War on behalf of the German Empire. Pursued by MI5 and awaiting his imminent arrest, Yu Tsun devises a plan to convey the military secrets he has uncovered to his German handlers. His plan is to conflate text with an actual space. Inspired by an ancient Chinese scholar, Yu Tsun imagines how a book about a labyrinthine garden might be made identical to an actual garden. So, just before he is arrested, he visits and murders a famous sinologist by the name of Albert. As he goes on trial for treason and murder, the Germans bomb the British secret artillery park at Albert, on the Somme. They have found out about the park's secret location from the British newspapers' reports about Yu Tsun's murder of 'Albert'. This is hypertexting par excellence: connecting two different ideas (or physical entities, or whatever) via a common word. In a sense, hypertexting is similar to a semantic collage of infinite size that is continuously constructed by free association, like a map of our collecting subconscious. One word may connect two or more disparate concepts together, which then proceed to connect via other common words with more concepts, and so on. Hypertexting makes the connections between concepts that are more than one degree of separation apart significant, for they are like knots in Ariadne's thread in the myth of Theseus. By following the thread one can navigate through the labyrinth of the World Wide Web: and that is exactly the principle upon which search engines function.

The discovery of the Internet, combined with the World Wide Web's system of information exchange, provided the missing ingredient necessary for the information revolution to take place. Unlike Otlet's centralised politics, the 1960s and 1970s counterculture movements were inspired by different, decentralised, even anarchic, political ideas. These ideas would exert a strong influence on the computer pioneers who replaced the hippies in San Francisco's Bay Area and the Valley. The civil rights movement in particular played an important role in recasting our approach to computers in peer-to-peer terms, where everyone was equal and had equal rights, where everyone was a producer and consumer of data at the same time – instead of approaching them in a centralised, hierarchical fashion. Psychedelic drugs also inspired a transcendental, and effectively Neoplatonic, narrative about computers and computer software. In a replay of the Eleusinian mysteries of ancient Athens, many citizens of the new computer *polis* of California rediscovered dualism through psychotropic drugs, and felt that consciousness could be projected beyond the human body. This decoupling of body and experience gave birth to technologies such as the video games, computer graphics and virtual reality; and informed transhumanist visions of the future, according to which humans and machines could potentially merge into cyborgs possessing augmented intelligence, shared consciousness and new types of sexuality. As the 1990s saw the birth of a new, unipolar world, in which the United States became the unchallenged superpower, technology caught up with the ancient narratives of love and fear of artificial beings. Thanks to the Internet, a new breed of digital deities emerged, which would quickly take control of the world. Ironically, they are called 'servers'.

A world ruled by servers

Technology author and MIT Professor Nicholas Negroponte was one of the original prophets of the digital revolution. In his classic book *Being Digital*,[19] published in 1995, he foresaw how digital media technologies would merge, and how computers and telecommunications would become one. More importantly perhaps, Negroponte predicted the global transformation from atoms to bits. Atoms, the essential building blocks of matter, make up tangible things such as CDs, books, newspapers and magazines. Bits are units of digital information. Nowadays, the music and publishing industries are amongst those that have witnessed, and suffered from, the digital transformation of their physical assets into bits. There are near-zero costs involved in copying things made of bits compared with copying things made of atoms. Bits can be transferred at the speed of light, and shared amongst billions of people over the Internet almost seamlessly. Atoms need complex logistics, trucks, truck drivers, retail stores and an army of people to work at them. In a battle of atoms against bits only a fool would bet on the former. Nearly twenty years after the publication of Negroponte's book, millions stream music on Spotify and Pandora, while only the diminishing few persist in buying CDs or vinyl. Newspapers and magazines strive to reinvent themselves in a digital age where content is mostly free. It is more likely that you are reading this book on your tablet, rather than holding a physical artefact made of paper. The law of digital transformation is simple: if something can be digitised it will be.

Another law that drives, and defines, our information age is the famous 'Moore's Law', named after Gordon Moore, the co-founder of Intel, who first identified it.[20] The law states that the power of computer technologies doubles every two years. Although it is more of an empirical observation rather than an actual law of nature, Moore's law fits very well with the data of computer evolution. Human ingenuity in devising successive innovations has contributed to the fulfilment of Moore's Law, the most notable innovation being our manufacturing ability to pack many transistors into a single silicon chip. In 1995 – the year Negroponte published his book – the state-of-the-art microprocessor[21] had 9.3 million transistors. Six years later the norm had risen to forty million transistors packed inside a chip. Today we are close to surpassing the fifteen billion transistors mark, which is very close to the ultimate threshold that nature permits before quantum phenomena kick in and render electronic gates useless.[22] Nevertheless, it is forecasted that, by 2020, molecular scale production will be used to go beyond the quantum threshold, and position each molecule individually on a chip. There are now two camps in the debate about when Moore's Law will actually collapse. The pessimists believe the ultimate limit will be reached in the next few decades. Optimists, such as Lawrence Krauss, predict that computers will continue to evolve exponentially for the next six centuries, with Krauss basing his calculations on the total information capacity of the universe.[23] This ultra-optimistic scenario suggests that the digital transformation that was set off on planet Earth in the 1990s has the potential to expand beyond the confines of our planet, our solar system and, indeed, our galaxy, and ultimately envelop the whole cosmos. This, as we saw, closely echoes the metaphysical predictions of Teillard de Chardin, an iconic figure amongst computer scientists and entrepreneurs. Krauss is effectively predicting a cosmic 'noosphere'.

Whatever the far future may bring, the fact is that, today, the smaller chips get the cheaper they become. This fall in price results in the further proliferation of small, cheap and very sophisticated mobile devices – such as smartphones, laptops and tablets – which can connect to the Internet wirelessly and provide us with a cornucopia of useful services and applications. The more consumers of the bits that digitise an ever-expanding variety of things previously made of atoms, the more investment in digitalisation and its relevant products is attracted – a virtuous cycle that lies at the heart of the contemporary information age. Nowadays, we consume data like never before in the history of the world. And Moore's Law seems to apply to data, too. It is estimated that there were 2.7 zettabytes[24] (or 2.7 sextillion bytes) in the world in 2012, a 50 per cent increase from a year earlier. Cisco, a manufacturer of router technologies, predicts that global Internet traffic will reach 1.6 zettabytes by 2018.[25] That's the equivalent of 250 billion DVDs of information:[26] the equivalent of all the movies ever made will cross global Internet networks every three minutes.

All these data are stored in vast computers called servers. As consumers demand ever smarter, leaner, lighter and 'cooler' machines to use and play with, the really hard work of storing, moving and manipulating enormous amounts of data is moved to the 'back end' of the Internet. This has been made possible thanks to so-called 'client-server' architectures. A 'client' is a smaller machine that exchanges information and accesses processing power from the more powerful 'server'. Whenever we access the Internet we connect our client machines (e.g. our smartphones) with a remote server. This server then connects with other remote servers around the world in order to return to our hand-held client whatever information he or she is looking for. The huge complexity of the Internet is invisible to the vast majority of us. As cloud technologies push physical servers out of small and big businesses, data and programs are increasingly stored in ever more concentrated server farms run by mega-companies such as Amazon. This transformation of moving complexity out of sight has many advantages. For businesses, it means dramatically lowering the cost of maintaining expensive servers on their premises. For consumers, it means not needing to worry about how information is processed or where it is stored, leaving them free to focus on enjoying the newfound bounty of information processing.

There are some interesting parallels between this technological trend and H. G. Wells's novella *The Time Machine*. This tells the story of a time traveller who visits Earth in the far future, only to discover that humanity has split into two separate races. The leisured Eloi are small, childlike adults who do little or no work, and who appear to enjoy the spoils of an apparently affluent society. But deep underground lurk the Morlocks, ape-like troglodytes who work the machinery and industry that make the terrestrial utopia possible. The downside of this seemingly harmonious arrangement is that the Morlocks have an unnerving habit of surfacing during the night to eat the Eloi. Although the computer servers that do all the hard work behind the information revolution may not physically eat us, like the Morlocks, our increasing dependency on them is definitely a matter of concern. The first warning that something was not exactly right with our increasing dependency on computers came only ten years after the invention of the World Wide Web.

As clocks ticked down the last seconds of 31 December 1999, many believed that the world was witnessing the countdown to an apocalypse. The 'Y2K bug' had become an obsession for many during the last few years of the twentieth century. Their anxiety had its roots in the legacy of computer systems programmed in previous decades, which did not account for the fact that, starting in the year 2000, counting years would have to be zeroed. It had been the traditional practice amongst programmers to note years with two digit numbers, and to code calculations on that basis. As the year 2000 approached, alarming reports began to circulate that resulting miscalculations could cause whole systems to go haywire. Experts talked about blackouts, water outages, and even nuclear weapons being fired by mistake. According to some reports,[27] more than US$300 billion were spent worldwide in order to provide fixes for the 'millennium bug'. In the end nothing happened. But no one was really certain about the outcome until the safe passing of the so-called 'event horizon' of midnight on 31 December, 1999.

Y2K was the first time that computer science borrowed terminology from quantum cosmology. An 'event horizon' describes a boundary separating our world of classical Newtonian physics from the unknown consequences of falling into a black hole. Crossing the event horizon, one falls into a 'singularity', which basically means a place and a time where the known laws of nature cease to hold. Even after the year 2000, apocalyptic scenarios involving the crossing of an 'event horizon' and the arrival at a 'singularity' would come to the fore again, this time substituting 'Y2K' with 'AI'. This is the infamous 'AI Singularity', which warns of the unpredictable and potentially disastrous effects of true Artificial Intelligence actually appearing (and to which I will return in more detail in the next chapter).

Ten years after the feared Y2K crisis, another incident reminded us how the dependence of humanity's vital institutions on automated computer technology can potentially cause a global catastrophe. Early in the afternoon of Thursday, 6 May 2010 the Dow Jones Industrial Average fell by 6 per cent in a matter of minutes. Not only that, but all kinds of crazy things started happening with stock prices: some fell as low as one cent and others shot through the roof at US$100,000 apiece with no obvious cause. In fifteen nail-biting minutes almost US$1 trillion of market capitalisation was wiped out. Yet five minutes later the Dow was back to normal, as if nothing had happened. The incident became known as the 'Flash Crash'.[28] The causes of it are still highly contested. The official explanation by the Securities and Exchange Commission blames a single badly timed and overly large stock sale. But this is disputed by many experts, who instead point at a set of financial computer technologies called 'high-frequency trading' as the true culprit. Basically, what these technologies do is exploit tiny, nanosecond-scale intervals between the placement of an order to buy or sell stock and the actual transaction. It is a technology that subverts one of the fundamental assumptions of a free market: that every player should have access to the same information. Those who have access to this technology make fortunes; those who do not are losers. The ethical downside of high frequency trading is further compounded by the obvious, demonstrable risks as regards the stability of global stock markets when fast algorithms take massive split-second buying decisions.

Despite the fact that complex software and hardware already take autonomous decisions that may have adverse effects on a global scale, there should be no desire to halt progress – as long as we understand the risks. Every day, the information age delivers value across every sector of our society. The digitisation of just about everything creates new opportunities for wealth and for finding fresh ways to solve problems across the whole spectrum of the human condition. Thanks to digital data and ever-accelerating computer power we are at the cusp of an era in which we can gain unprecedented insights into natural phenomena, the human body, markets, Earth's climate, ecosystems, energy grids, and just about everything in between. Norbert Wiener's cybernetic dream is slowly becoming a reality: the more information we have about systems, the more control we can exercise over them with the help of our computers. Big data are our newfound economic bounty.

The big data economy

In 2010, I took a contract as External Relations Officer at the European Bioinformatics Institute (EBI) at Hinxton, Cambridge. The Institute is part of the intergovernmental European Molecular Biology Laboratory, and its core mission is to provide an infrastructure for the storage and manipulation of biological data. This is the data that researchers in the life sciences produce every day, including information about the genes of humans and of other species, chemical molecules that might provide the basis for new therapies, proteins, and also about research findings in general. These data represent an absolute gold mine for biology. Healthcare, food production and security, environmental protection and energy are just a handful of the industries that benefit from research using biological data. And these data are exploding. Take, for example, data about the human genome. As the cost of sequencing human DNA drops, terabytes of genome data need to be stored; all this information is very valuable as researchers try to unlock the connection between genetics and disease. At the time that I worked for them, EBI's challenge was to increase the capacity of its infrastructure in order to accommodate this 'data deluge'. As someone who facilitated communications between the Institute and potential government funders across Europe, I had first-hand experience of the importance that governments placed on biological data. Almost everyone understood the potential for driving innovation through this data, and was ready to support the expansion of Europe's bioinformatics infrastructure, even as Europe was going through the Great Recession. The message was simple and clear: whoever owned the data owned the future.

Governments and scientists are not the only ones to have jumped on the bandwagon of big data. The advent of social media and Google Search has transformed the marketing operations of almost every business in the world, big and small. Tools have been developed to 'mine' the text written by billions of people on Facebook and Twitter, in order to measure sentiment and target consumers with, hopefully, the right products. As we leave more data about ourselves on the Web, companies exploit this to identify not only who might be their best customers, but also at what time of the day and under which circumstances we are most likely to buy their products and services. This unprecedented insight into our personal lives by government as well as private companies can easily backfire. In 2012, it was revealed that the American retail giant Target had been using an algorithm designed to predict if a woman was pregnant according to what shopping she did. The purpose of the algorithm was so that other products specifically relating to pregnancy could be suggested to the woman, thus increasing her spend. However, Target's marketing strategy backfired when a teenager was sent several coupons advertising baby-related products by mail to her home address. Neither she nor her parents were aware she was pregnant[29] at the time. Target's algorithm knew it before they did! The incident was picked up by the media, and created the spooky feeling that Target was actually stalking its customers, gravely affecting the company's reputation.

That spooky feeling that someone is watching us online was further accentuated following the revelations of US National Security Agency (NSA) contractor Edward Snowden in 2013. According to classified documents leaked by Snowden to the *Guardian* newspaper, the NSA spied on US citizens as well as on citizens from other countries, including top foreign politicians, by 'listening in' to their conversations over the Internet. Data from these conversations were stored in massive computer server farms, where they were mined by algorithms searching for patterns. Although the NSA was authorised by the US administration to execute such a global surveillance in order to prevent terrorist attacks, the fact that the US government spied on its own citizens created a dangerous precedent that struck at the foundations of a liberal society and of the US Constitution. Spying on allies also smacked of industrial espionage, making the pretext of defence of the realm appear paper-thin. Snowden's revelations brought home the more general message that our data on the Internet can be collected and manipulated by anyone who may wish to do so, for whatever purpose they want. When we chat absentmindedly on Facebook with friends, or hit the 'Like' button, as we pass from one Web page to the next and click on articles, we leave behind a trace of data – our 'digital scent' – that can be picked up and used to find out things about us. The Internet has opened a window to our personal lives, feelings, thoughts, relationships and aspirations, even our vices. And we do not, and cannot, know who is watching us at any time.

The interconnectedness of people and machines resulting from the computer revolution of the mid-twentieth century has made our world more complex than ever before. Financial systems, energy grids, defence systems, transportation – just about everything – have undergone or are undergoing digital transformation. Everything is becoming digitised as data are increasingly manipulated by logic algorithms on remote servers. Yet engineers know well that complexity is synonymous with instability. Computer viruses are nowadays a menace for computer systems that may seem somewhat remote from everyday life, but imagine the day when your car will be run mostly by software. What if someone infected your car with a virus that could make it spin out of control when it reached eighty miles per hour? And what about all the smart things with which we are all beginning to furnish our homes: intelligent thermostats, fridges that order food when it runs out, telemedicine devices that monitor our health? The evangelists of the 'internet of things' proclaim that our lives will be simpler and more productive when the things we use can take decisions on our behalf. This is happening already, but will explode in the next few years. According to Cisco CEO John Chambers there are some thirteen billion devices connected to the Internet today, a number predicted to grow to fifty billion by 2020, and 500 billion by 2030.[30] The Internet of things will result in US$19 trillion in profits and cost savings in the private and public sector, and will be ten to fifteen times larger than the Internet today in terms of number of connections.

Things that think, talk and do

The 'Internet of things' is postmodernism reinventing panpsychism – the idea that all things share a mind, or a soul. Platonic ideas where the pattern or form is privileged over matter, so prominent in computer science given the software/hardware paradigm, are well on their way to invading every aspect of our daily lives through the chips implanted in household devices; in farmland where they are used to track water and fertilisation levels; floating in the air to monitor pollution; and soon wearable sensors embedded in our clothes to track and report our vital signs. We are already talking to our computers, asking them questions, requesting they book an appointment in our calendar or a table at a restaurant. Soon we will be talking to our homes, our cars, our furniture – and receiving a reply.

But have we stopped to consider the vulnerability of such embedded computer systems to computer viruses, spies, terrorists, pranksters, and whoever might wish to access our data for nefarious goals? And what about policing? Unlike the relatively benign case of the teenage mother-to-be and Target, there are some serious implications. For instance, the police of the future might attempt to prevent crime simply by cross-referencing the behavioural data of citizens, as in the film *Minority Report* (2002). Do we really want to allow this? And what about politicians, or civil servants: one has to ponder their usefulness in a world where interconnected computers with access to enormous amounts of data can take much better decisions on policy for a tiny fraction of the cost of a human government. What would the big data economy and the Internet of things mean for politics and democracy? These are questions that will increasingly become pertinent to the debate about our future, and which I will discuss at the end of the book. But what about the technology itself? If the increased complexity of computer systems means increased insecurity and unwanted interdependency, what can we do about that?

There are two ways to deal with the problem of computers running human affairs. One is to make things simpler. Some of us may aspire to go 'off grid', simplifying our lives by going back to nature, throwing away our smartphones or unplugging from the Internet. I suspect, however, that only a few would choose that path, and those who did would soon discover the huge challenges of practically cutting themselves off from the civilised world. For better or for worse, humanity can only move forward and deal with such complexities by using the only realistic means available. Until today, those means are human operators overseeing the machines. Take, for example, human air traffic controllers who still oversee computers. Nevertheless, the option of humans as the ultimate overlords and supreme authority over the machines is increasingly becoming impractical if not downright impossible. The exponentially increasing power, complexity and interconnectedness of computers are more than any human, or collection of humans, can comprehend, let alone control. Our only other option is therefore adding yet another layer of complexity, of a non-human kind. What if we had super machines that could watch over us? Machines that monitored other machines and ensured no one spied on our data, machines that defended our vital computer systems and corrected the instabilities whenever they might occur, that guaranteed there could be no more 'millennium bugs' or 'flash crashes' or a digital apocalypse? What if we had machines that were truly intelligent and that would be our guardians?

15
MACHINES THAT THINK

We have come a long way since Aristotle had the insight that logic follows rules. We saw how Boole and Frege pushed this insight further by codifying logic, thus enabling the development of computer languages that code logical rules. In the fullness of time, a torrent of inventions and innovations – such as the electric bulb, electromechanical relays, the transistor and miniaturisation – facilitated the development of advanced electronics. Claude Shannon showed that logical rules could be executed using electronics, and Alan Turing, together with John von Neumann, demonstrated how to build electronic machines that solved (almost) any logical problem. And that was how the modern digital computer was born – a quadrillion times faster and more powerful than its steampunk ancestor the emblematic Analytical Engine. Nevertheless, Babbage's big idea of separating hardware from software, and thus creating a general-purpose machine that could run any set of instructions, was preserved in time as the foundation of modern computer architectures. Indeed, it is the separation of hardware from software that allows the tremendous advances in computer engineering, and facilitates the exponential evolution of machines, at a rate that sees them becoming twice as powerful every eighteen months.

Powerful computing machines interconnected over the Internet, coupled with the near-zero cost of transmitting and copying digital information, drive a global trend for digital transformation. We are nowadays the denizens of a digital noosphere: creators, consumers and manipulators of vast amounts of digital data. The deluge of big data that comes from the digitisation of almost everything, and the value for businesses and governments that these data encapsulate, are taking the world economy into a new era increasingly called 'the second machine age'.[1] The 'first age' occurred when the invention of the steam engine multiplied humanity's capacity for manual labour. In the 'second age' the computer multiplies our capacity for mental labour. As computers increasingly become more 'intelligent', they are bound to transcend their current number-crunching duties and take over jobs traditionally associated with human, white-collar workers. All the signs point in that direction. Within the past two years Google, one of the biggest companies in the computer industry,[2] acquired a number of companies in Artificial Intelligence and advanced robotics. Facebook also announced that one of the most prominent AI researchers in the world, Professor Yann LeCun of NYU's Center for Data Science, would be joining the company to direct a massive new AI effort. These global companies move towards smarter machine technologies because they understand the challenges and opportunities entailed in owning big data. They also understand that it is not enough to own the data. The real game changer lies in understanding the data's true significance.

Take, for instance, Professor LeCun, a pioneer in developing deep learning algorithms that can interpret meanings and contexts of symbols and images. This technology is valuable for Facebook as it aspires to increase the ways in which it serves its billions of customers – and the advertising industry – by extracting meaning from its colossal and ever-expanding archive of user-generated content. Google has a similar aspiration: it wants to use AI technology to understand context and meaning, and thus provide better search resources, video recognition, speech recognition and translation, increased security, and smarter services when it comes to Google's social networks and e-commerce platforms. When Google spent half a billion dollars to acquire the British company Deep Mind, it was in fact hedging a bet that Artificial Intelligence will define the second machine age.

In this chapter I shall explore what all this means. How close are we to truly intelligent machines – complete with self-awareness? What will the repercussions be for our economy and society as thinking machines begin to replace us in the workplace? Are we in danger of extinction from thinking machines that will one day become self-aware and take over the world – making the millennium bug and the flash crash incidents seem like child's play? How close are we to the notorious 'AI Singularity'?

The wise men of Dartmouth

Artificial Intelligence, as a distinct scientific discipline, was born in the summer of 1956 during a conference on the campus of Dartmouth College in New Hampshire. It was a truly historical event, and those who attended would go on to contribute major innovations in the field of AI in the years to come. The principal inspiration for the conference was Walter Pitts and Warren McCulloch's demonstration of the equivalence between a biological neuron and a logical function. One of McCulloch's students was the young Marvin Minsky, who would build the first electronic neural net. He was also one of the main organisers of the Dartmouth conference. Other notable organisers were John McCarthy,[3] cybernetics giant Claude Shannon and computer pioneer Nathaniel Rochester.[4]

The late 1950s was a period of scientific exuberance during which new ideas received ample funding from defence budgets. McCulloch and Pitts' discovery suggested that a logical machine could imitate a brain and, ultimately, attain sentience. The founders of AI decided to part ways with general cybernetics, and focus on the neuron equivalence. Their goal was to program computers to perform human mental functions such as learning, solving logical problems and communicating using natural language. The stakes of AI research were high. An intelligent computer, which was capable of accessing boundless information and processing it millions of times faster than the smartest human, could potentially solve every problem, including how to run an economy more successfully, win every possible battle or develop new weapons. The global power that possessed such a technology would rule the world, a notion that did not go amiss with government funders during the height of the Cold War. Almost everybody saw the development of an artificial mind as inevitable. In 1968, Arthur C. Clarke and Stanley Kubrick imagined HAL 9000, a computer so humanly intelligent that it could go mad.[5] Their forecasted date for the existence of true Artificial Intelligence was the year 2001. With the benefit of hindsight, we know that those initial predictions were overoptimistic. But in the context of the time that they were made they seemed reasonable. The claims of pioneering AI research were founded on the premise that logic was the active ingredient of intelligence. There was a strong cultural element at play in this claim; the essence of 'humanness' was assumed to be about the ability to reason and do clever things such as solve complex logical problems. Emotions were ignored as part of a lower, and rather uninteresting, 'animal' or 'primitive' aspect of being human. Focusing on logic, the AI pioneers suggested that the advent of artificial intelligence was a matter of scale. As computers became better at performing so their intelligence would increase until it reached, and surpassed, that of humans.

The pioneers of AI explored many ideas including using algorithms for solving general logical problems, or simulating parts of the brain using artificial neural nets. And although they produced some very capable systems, none of them could arguably be called intelligent. Of course, how one defines intelligence is also crucial. For the pioneers of AI, 'artificial intelligence' was nothing less than the artificial equivalent of human intelligence, a position nowadays referred to as 'strong AI'. An intelligent machine ought to be one that possessed general intelligence, just like a human. This meant that the machine ought to be able to solve any problem using first principles and experience derived from learning. Early models of general-solving were built, but could not scale up. Systems could solve one general problem but not *any* general problem.[6] Algorithms that searched data in order to make general inferences failed quickly because of something called 'combinatorial explosion': there were simply too many interrelated parameters and variables to calculate after a number of steps. An approach called 'heuristics' tried to solve the combinatorial explosion problem by 'pruning' branches off the tree of the search executed by any given algorithm; but even this was shown to be of limited value. In the end, AI researchers came to realise that problems such as the recognition of faces or objects required 'common sense' reasoning, which was fiendishly difficult to code. Given the limitations of computer technology in the 1950s, 1960s and 1970s, they surmised that what was lacking were more powerful computers.

Alas, they could not see the big elephant in the room, which was symbolic logic itself. Regardless of what Aristotle, Boole, Frege and Wittgenstein might have said or proved, there were far too many things in the world that lay beyond logic, yet were very much part of life and experience. General-purpose computing did not translate directly to general intelligence. Although they hated to admit it, early AI researchers had actually discovered that general intelligence was probably impossible to code in any computer language.

Take, for example, simple concepts we humans have for everyday objects such as 'chair' or 'restaurant'. We only need to see any one chair once, and will thereafter recognise any other 'chair' we come across as such, regardless of how different it looks. In fact, we might even say that something 'looks like a chair' even if it is not actually a chair. This kind of general reasoning, so straightforward to us humans, proved extremely difficult to code using logic.[7] Could it be that logic did not apply to everything after all? Were there mental processes in the human brain that were, somehow, 'illogical'?

At the microscopic level of the neuron, processes could be simulated using Boolean logic, just as Pitts and McCulloch had demonstrated. But as neurons grouped and clustered together, and organised themselves into ever more intricate and higher levels of complexity, something curious happened in the human brain that AI research could not reproduce in a computer. Perhaps general intelligence and self-awareness were functions that no Turing machine could possibly emulate. It was an exasperating and tormenting thought that led to many getting cold feet about the future of AI.

Not surprisingly, by 1974 the initial enthusiasm with AI had blown off,[8] and was replaced by disappointment and often ridicule. AI systems did a few clever things, but they were a long way from earning the epithet 'intelligent'. The generous funding from defence budgets ended and the long 'winter of AI' set in. As computers continued to evolve apace in the business world, taking over more and more everyday tasks, artificial intelligence seemed irrelevant, ineffective and rather quaint.

Back in from the cold

AI was resurrected from the dead in the early 1980s by two seminal events. Firstly, Japan announced a multimillion-dollar investment in '5th generation computing' that aimed to transform computers into intelligent machines that could reason like human beings. This marked a fresh attempt to achieve the original goal of producing general intelligence in machines: to make computers that could understand language, comprehend images and carry on conversations. The fact that the up-and-coming technological giant of Asia was willing to invest considerable resources[9] in a big science project focused on Artificial Intelligence came as a wake-up call for Western governments. Could it be that the Japanese saw a golden opportunity in an area where the West had thrown in the towel too soon?

Meanwhile, in the Western countries where AI was originally born, the definition of the field had changed. Because of the original failure to deliver the goods, the scope of AI research had become narrower. Those early ambitious dreams of self-aware machines were over, and now a more timid generation of researchers discussed the practical applications of symbolic logic to address problems that traditional computer scientists found difficult to solve. This represented a major redirection of AI research, the legacy of which is still with us today. Although the term 'Artificial Intelligence' has been kept in use, its meaning has subtly changed. From the 1980s onwards, 'AI' came to mean computers performing tasks normally considered 'human' activities, such as taking decisions on the basis of inexact data, or understanding natural language. The field no longer claims that a computer needs to be 'intelligent' in any intrinsic, philosophical or general way, but that computers that have been programmed effectively can solve problems by applying reasoning in certain specific application areas. From general-purpose, AI has become purpose-specific. However, this tectonic shift in meaning still creates much confusion in the media and among the general public today. Most ordinarily people and non-science journalists still think of AI as computers becoming as intelligent as humans. But this is not what actually takes place in modern AI labs. What researchers there try to do is to produce software and hardware that would work together in such a way for a computer to be able to perform human-like tasks better, more efficiently, in a manner less error-prone and a lot more quickly. For this to happen machine self-awareness is not a prerequisite.

Nevertheless, this semantic shift made AI more popular with funders seeking actual results, and more successful in its commercialisation. The 1980s saw the rise of a type of AI programs called 'expert systems'. Based on logical rules derived from the knowledge of experts, these systems could answer questions and solve problems in very specific domains of knowledge. The earliest expert system was 'Dendral' developed at Stanford University by Edward Feigenbaum in 1965. The 'knowledge base' of this expert system was made up of rules derived by experts in the field of spectrometry, the science of guessing what a chemical compound is made of by shining light through the compound and analysing the spectrum of the light that the compound reflects back. This is a highly technical job that needs not just good data but also good scientists who can interpret the data according to their long experience. Dendral could identify the chemical composition of compounds when fed measurement data from a spectrometer, just like a human expert would. From the same laboratory at Stanford came another landmark expert system, MYCIN.[10] Built in 1972, the system could identify certain bacteria that caused infections and then recommend antibiotics, taking into account patient data such as their body weight.

The expert systems of the 1980s took inspiration from Dendral and MYCIN, and became popular because they offered technological solutions to many business problems. Management gurus realised that 'expert knowledge' was an important driver of commercial success and innovation. The catchphrase 'knowledge capital' was used to signify the collective expert knowledge of a company. Information technology systems would increasingly be used to store and access this form of knowledge, and 'knowledge engineering' became a field in itself. In this context, expert systems were quickly developed and adopted across many businesses, including finance, banking and health industries. One of the reasons that expert systems began to proliferate in mainstream computing was their ability to draw inferences from uncertain or incomplete data. Let me delve a little deeper into how this was made possible, by telling you a little more about my research.

The expert system that I developed as part of my doctoral thesis in 1989 is an illustration of a computer executing 'uncertainty reasoning'. The system combined a knowledge base of several dozens of rules extracted from interviews with human medical doctors, with an 'inference engine', which was another set of rules that described how this knowledge was to be manipulated.[11] The application lay in determining the status of patients in an intensive care unit by measuring their blood gases, and by assessing their general clinical status and medical history. The expert system took as input the results of a blood gas analysis, then put a number of questions to a human user who responded using 'yes' or 'no', and finally produced a series of possible diagnoses ranked according to probability. Uncertainty is the norm in medical diagnosis, and that is how human doctors think when trying to weigh the evidence of a patient case, a process called 'differential diagnosis'. In order for my system to determine which of the possible diagnoses was the more probable, I borrowed ideas from fuzzy set theory and used a statistical technique called Bayesian inference. The latter is a very common tool used to make decisions about something whose nature is uncertain but for which you have a series of information updates. If you do not know what that uncertain something actually is, you build a number of hypotheses around it. As you receive more information updates, your relative belief in your various hypotheses changes. Bayesian inference manipulates increasing information data as 'additional evidence'. The hypothesis that gets the highest 'score' is the most 'probable'.

Bayesian inference is the closest that statistics has to offer to subjective probability. But is this the means actually used by human brains to evaluate what outcome is the more probable? Given what we know so far from neuroscience, the answer seems to be no. Expert systems execute symbolic processing by manipulating symbols on the basis of logical rules. Mental processing is not symbolic but biological. However, this ontological discrepancy between Artificial Intelligence and natural intelligence was irrelevant in the 1980s, given that 'AI' had already changed direction. By then, no serious AI researcher would admit to wanting to reproduce a human mind in a machine. AI was as good as it was effective at solving real-world problems, such as accessing vast knowledge bases and coming up with a hierarchy of logical hypotheses and answers. Nevertheless, this ontological discrepancy was quickly forgotten as computers followed Moore's Law and became more powerful; as they began to exhibit powers that people identified with 'real intelligence'.

On 11 May 1997, 'Deep Blue', a chess-playing computer developed by IBM, won a six-game match against world champion Garry Kasparov.[12] This was the first time that brute-force computing demonstrated how easily it could be mistaken for 'superior intelligence'. Deep Blue was designed specifically for chess. Its massively parallel hardware used thirty top-performance microprocessors that explored 200 million positions per second.[13] Its database stored hundreds of thousands of master games that were used to evaluate which moves the computer should make. And yet it was probably a bug in the system that caused the world's human champion to feel anxious and lose his concentration. Towards the end of the first game Deep Blue, unable to select a move, defaulted to a last resort fail-safe and picked a move completely at random.[14] Kasparov misunderstood the move for creativity, lost his concentration and subsequently lost the game. It was an ironic instance of the Turing Test succeeding in fooling a human being into believing that a computer was really intelligent! Despite Kasparov's protestations, IBM did not offer him the rematch he demanded, and soon after Big Blue was dismantled. Nevertheless, the historical event of a computer beating the world champion of chess is considered a watershed in the evolution of Artificial Intelligence. IBM's stock price spiked, and the company prepared for its next big move in AI.

Meanwhile, the defence supremos who had pulled the plug on AI research back in 1974 started thinking about it once again. In 2004, the US Defense Advanced Research Projects Agency announced a US$1 million prize for the construction of a self-driving vehicle capable of navigating a 150-mile uncharted route. None of the robot vehicles that took part finished the route that year. The furthest any of the contestants managed was 7.3 miles. Next year, DARPA repeated the challenge. This time five vehicles completed the course, the team from Stanford University gaining first place. Since then DARPA has repeated the robotics challenge, to include autonomous vehicles capable of finding their way in an urban environment, as well as humanoid robots. The contestants have consistently produced better products over the years. This rapid evolution in performance is very telling of how quickly engineers can integrate new systems nowadays, and innovate. Google and others are currently developing prototype commercial driverless cars, which we should expect to become part of our everyday lives by the next decade.

The other seminal event that signalled that something big was changing in the field of Artificial Intelligence took place in February 2011, and was televised. Watson – another computer developed by IBM – beat two former, human, winners of the popular American TV quiz *Jeopardy!* and won the prize of a million dollars. Watson was a truly amazing machine. It was not a singular entity but a cluster of ninety servers, each one equipped with multiple processors. Its massively parallel hardware architecture was capable of supporting millions of searches into its knowledge base. For the purpose of the TV quiz, the engineers at IBM loaded Watson with 200 million pages of data, including dictionaries, encyclopaedias and literary articles. Moreover, Watson communicated in natural language. You asked it a question, it understood it, and returned an answer. For this to happen, Watson's designers exploited the whole arsenal of AI tools and techniques, including machine learning, natural language processing and knowledge representation. What the success of their creation demonstrated was that brute computing force could overcome the obstacles that the AI pioneers faced in the 1960s and early 1970s. Bigger, stronger, faster were very meaningful words when it came to increasing machine intelligence.

Deep Blue, DARPA's navigational challenge and Watson ushered AI to the fore of public awareness and debate. Their success challenged expectations with regards to computers. AI machines were not simply processors of data, but demonstrated capabilities hitherto considered uniquely human. To be able to beat the brilliant mind of the world chess champion, to drive a car, to understand language including metaphor and slang, were harbingers of more things to come. But what might those things be?

In 2013, IBM announced the release of an application programming interface (API) for Watson. Using this API, software developers can integrate the natural language and knowledge search capabilities of Watson in order to build new applications and services. This means that a very important, ground-breaking program of AI effectively becomes a 'machine element', i.e. an elementary component (similar to a mechanical screw, or an electrical resistance) with which an engineer can build a more complex machine. With the ability to integrate sophisticated machine elements in new software products and designs, with computing power doubling every eighteen months, and with huge economic rewards for those who innovate successfully, we have entered a new, accelerated phase of technological development. This isn't simply about a new generation of computers doing things faster and more efficiently. Artificial Intelligence machines are becoming increasingly capable of outsmarting humans in almost every aspect of applied intelligence, of learning faster than we do, and of communicating with us in our own languages. Without doubt, our world is entering uncharted waters. What will the repercussions of these new intelligent machines be, as they take over tasks that intelligent, well-educated people have done for centuries? Can we imagine a computer acting as our lawyer, or our doctor? And what about machines capable of designing and building other machines? Could intelligent machines become so powerful and clever that they could outmanoeuvre, and ultimately replace, their human creators? Is our technology on the way to making us obsolete?

The second machine age

As the global big-data economy expands manyfold thanks to the Internet of things, a new generation of powerful Artificial Intelligence applications has arrived that is capable of further enhancing the capabilities of computer systems. We are truly on the cusp of one of the most significant moments in human history: the beginning of a new industrial revolution. But there is also a downside. An Oxford University research paper estimates that 47 per cent of our current occupations are at risk of becoming automated over the course of the next few decades.[15] This represents a prognosis of monumental proportions. It suggests that almost half of the professions from which people earn a living today will be stamped out by mid-century, because intelligent machines will be doing those jobs better, faster and more economically.

Unlike the previous machine age of the first Industrial Revolution, the next one will not threaten manual blue-collar jobs, but those of highly paid, expert, white-collar workers. Doctors, lawyers, engineers, accountants, managers, designers, architects, are forecasted to become victims of computer automation. According to American economist Tyler Cowen, in the near future only an elite 10–15 per cent of the working population will have the intellectual capacity to master tomorrow's AI technology, and that will make them very rich indeed.[16] The rest of us will have to make do with low incomes and rather unfulfilling lives, or – at best – work as service providers to the rich. It is a very bleak vision of the future, yet the economic trends of recent decades seem to support it.

Since the early 1970s, a sustained discrepancy has existed in developed economies between productivity and median wages. Although productivity, measured by output per worker, has been constantly rising, median wages have remained nearly flat. In particular, between 1973 and 2011 the median hourly wage in the USA grew by just 0.1 per cent per year, while productivity in the same period grew by an average of 1.56 per cent.[17] Increased productivity means more 'bounty', or wealth to be shared, as the economic pie grows bigger. But barely increasing median wages mean that this newly produced bounty is not shared fairly. The 'spread' of wealth between workers and capital owners has widened between 1973 and 2011. This means that most people in the world have not enjoyed any material benefits from the increase in the 'bounty' of the information age, because their incomes have remained low. High 'spread' in income also suggests that the bounty – or the new wealth created by the information age – has been mostly channelled to the very rich. Indeed, according to research by French economist Thomas Piketty, we now live in an age similar to the pre-industrial times in that a small minority of people – the notorious 1 per cent – owns most of the world's wealth.[18] The discrepancy between an increased 'bounty' and a widening 'spread' is likely to get much worse as we move further into the second machine age. And that's because automation and Artificial Intelligence systems will create more wealth but also obliterate many jobs.

From the 1960s, when automatic and numerical controls started to take over assembly line jobs in the manufacturing industries, to today's digital technologies, factories and companies have increasingly been able to produce more and better products with less manual labour. Even in China, where labour costs are rising and losing their global competitiveness, factory bosses today are beginning to commission armies of industrial robots to take over. Foxconn, a manufacturer of electronics and gaming consoles, made the news in 2012 by announcing it will replace a million workers with a million robots (aptly named 'Foxbots'). According to the Frankfurt-based International Federation of Robotics, China will become the biggest consumer of industrial robots by 2014. Chinese factories will thus continue to increase their productivity and the quality of their products at a lower cost. It makes perfect sense if you are the CEO or the owner of the factory, despite the fact that your economic astuteness will put a good many people out of work. More robots and more AI will push ever more people out of steady employment around the world. What will the impact of this be on societies? It helps to look at what happened in the past, when something very similar took place.

Between 1811 and 1816, Britain was shaken by massive riots with workers protesting against Parliament's decision to revoke a 1551 law that prohibited the use of gig mills in the wool-finishing trade. The legendary, and probably fictitious, leader of the movement was called Ned Ludd. He lent his name to the word 'Luddite', which has come to mean anyone who resists technological progress. As such, the word has a rather negative connotation in our own time. That is because it has been shown that technological change is socially and economically beneficial in the medium to long term. Despite the short-term rise in unemployment caused by the mechanisation of labour, in the early nineteenth century Britain's living standards rose considerably within two generations. In the Victorian age, Britain became one of the wealthiest nations in the world. Automation might initially put some people out of work, but it also creates new jobs in the medium to long term that are better paid and more interesting. Therefore, the 'bounty' brought by new technology does not only entail a degree of income inequality, but ultimately brings more prosperity to all. At the end of the day, it does not matter how much richer our neighbour becomes, as long as we become better off as well. Some income inequality is in fact socially useful, because it gives incentives to capable people to do their best, be innovative and take risks. Inequality becomes socially intolerable only when the bounty does not filter through. If we remain poor while our neighbour becomes a lot richer every year, then it is only human to begin to feel that inequality is a synonym for injustice.

Technophile optimists suggest that the second machine age will increase the bounty manyfold, and that the lives of our children and grandchildren will be the better for it. That they will enjoy longer and healthier lives thanks to, for example, intelligent machines capable of processing genetic data and discovering new and cheaper therapies. Artificial Intelligence computers will be able to accelerate technological innovation in all other areas as well, including renewable energy, food security, environmental conservation and space exploration. At the other end of the scale, and if predictions like Cowen's come true, our offspring will have limited employment and low, or erratic, incomes. These conflicting forecasts polarise the debate about the future of our developed, democratic societies. Economists on the left, including Piketty and Nobel laureate Paul Krugman,[19] propose that a larger state should tax the super-rich in order to provide social safety nets for a future, mostly jobless, middle class. Libertarians and conservatives counter these arguments with the suggestion that increased dependency on a nanny state will ruin the moral fabric of society.

But perhaps this political debate is missing the point by failing to consider the most critical factor of the second machine age – the fact that Artificial Intelligence is a technology unlike any other. Connected to colossal amounts of data and knowledge, with unlimited access to billions of smart devices that regulate almost every aspect of human life, AI systems have the potential to become the ultimate controllers of everything. From being our servers, intelligent computers may become our masters. The impact of Artificial Intelligence in society will thus be enormous, and rather unpredictable. It may not merely necessitate the redistribution of income but a radical reinvention of our political systems. Indeed, some go way beyond that. They warn that a super-intelligent AI will threaten the very survival of the human species.[20]

AI Apocalypse

In 2014, Max Tegmark, a prominent physicist at MIT, wrote in an op-ed in the *Huffington Post*[21] that he is in no doubt that one day computers will beat humans at all tasks and develop superhuman intelligence. After that point, he claimed, everything on Earth will change. Machines will outsmart the markets, outinvent and outpatent all human researchers, and outmanipulate all human leaders. In a follow-up public letter printed in the British newspaper the *Independent*, co-signed by Stephen Hawking, computer scientist Stuart Russell and physics Nobel-winner Frank Wilczek,[22] Tegmark and his peers raised the alarm about what might happen if AI takes over. Taking its lead from the film *Transcendence* (2014), these prominent scientists argued that the threat of human extinction is very real, very serious and closing in upon us. They are not the only ones worried about AI taking over the world. Ray Kurzweil – inventor, entrepreneur and currently the head of AI research for Google – thinks that this will happen by 2030. But how did all this talk about the AI Singularity start? The answer, not surprisingly perhaps, is to be found not in science but in science fiction.

Vernor Vinge is a computer scientist, science fiction writer and winner of the prestigious Hugo Award for science fiction. In his novels, particularly in *The Peace War* (1984) and in *Marooned in Realtime* (1986), Vinge was the first to explore a fictitious time in the future that he called 'the technological singularity'. This is when the human race has transcended into a different form of existence with the assistance of exponentially improving sentient technology. He expressed these narrative ideas more explicitly in a 1993 essay, arguing that the creation of superhuman Artificial Intelligence will mark a point in history where 'the human era will be ended'.[23] The main argument for the inevitability of the AI Singularity in Vinge's essay is Moore's Law. He writes: 'progress in computer hardware has followed an amazingly steady curve in the last few decades. Based largely on this trend, I believe that the creation of greater than human intelligence will occur during the next thirty years.'[24]

Ray Kurzweil adopted Vinge's argument in a series of popular science books that explore the technological drivers, and potentially devastating impact, of superhuman Artificial Intelligence. Kurzweil marks the year 2030 as a watershed by extrapolating, like Vinge, from today's exponential improvement of computers according to Moore's Law:[25] 2030 thus becomes the year that computer complexity will surpass the complexity of information processing in the human brain. Deep Blue, driverless cars crossing the Mojave Desert, and Watson beating humans at *Jeopardy!* all seem to validate the arguments made by Vinge and Kurzweil. Brute computer power has made computers more 'intelligent'. Nevertheless, underneath the correlation between powerful computing and intelligent behaviour lurk two fundamental assumptions that deserve closer examination.

The first assumption is that our computer technology, whose architecture is different from that of the human brain, is nevertheless capable of exhibiting every aspect of human intelligence, including self-awareness. The second assumption is that, as computer complexity increases by a factor of two every eighteen months, superhuman intelligence will somehow spontaneously emerge after computers reach the threshold of the brain's complexity.

I would like to call the first assumption 'the aeroplane metaphor'. Proponents of the AI Singularity hypothesis claim that computers do not have to be like human brains in order to exhibit superior intelligence, or in other words that they can be ontologically different and yet exhibit the same functionality. It would not be the first time that human engineering has surpassed nature, by inventing an ontologically different way of achieving the same end results. Aeroplanes fly differently from birds or insects, but fly nevertheless. Moreover, they fly faster, higher, longer, and carry many people inside them as well. The history of aviation has many similarities to the history of AI. Just like the early AI researchers of the 1950s, when people began to think about artificial flight they tried to emulate nature first. Leonardo da Vinci designed a human-powered 'ornithopter' that flapped its wings like a bird. Similarly, the pioneers of AI tried to reverse-engineer the human brain in order to build a machine that thought. It was only when engineers departed from faithfully mimicking nature that they got promising results. The first aeroplane built by the Wright brothers did not flap its wings, but glided using fixed wings and an aluminium engine. AI brushed off ridicule when it stopped trying to emulate real neurons and instead used powerful semiconductor processors and smart algorithms to parse natural language or beat Kasparov at chess. But do aeroplanes actually 'fly'?

If we wanted to be precise with our words we ought to say that aeroplanes glide. Gliding is one aspect of natural flying, but not the whole story. Aviation engineers used the word 'fly' for aeroplanes metaphorically in the early twentieth century, and as shorthand. It was also a good way to communicate the idea of machines taking to the sky like birds. The metaphor was used so frequently over time that it became a literal in our collective consciousness. We now think of aeroplanes as flying machines – just like we think of birds as flying biological machines.

The aeroplane metaphor is replicated in contemporary Artificial Intelligence. Having abandoned any claims to self-awareness, modern AI systems are called 'intelligent' metaphorically, not literarily. Metaphorically speaking, computers have already achieved superhuman 'intelligence', in the same way that aeroplanes have achieved superavian gliding. The confusion in the terms suggests that the aeroplane metaphor seriously undermines some of the arguments of the AI Singularity hypothesis. 'Superhuman intelligence' is not semantically equivalent to 'a computer possessing the whole spectrum of cognitive capabilities that a human brain has'. Computers supersede us only in specific subsets of intelligence. Brute computing power does not suffice for computers to achieve the whole spectrum of the human brain's cognitive abilities.

The second assumption underlying the AI Singularity hypothesis is that self-awareness will somehow 'emerge' from increasing computer complexity. Let us call this assumption 'pseudo-cybernetic' and examine it closer. We tend to associate intelligence with complexity because we are, apparently, very complex creatures. When we observe other life forms on Earth, the findings suggest that simpler organisms are less intelligent than us. Complexity and intelligence seem to be proportionally related. We also observe different kinds of intelligence in which complexity also seems to play a key role. Many insects, as well as funguses and microbes, exhibit swarm intelligence: although the individual creatures possess minimal information-processing capability, the colony as a whole exhibits an infinitely more refined capability to strategise and adapt. Many neuroscientists would argue that the human brain is nothing but a colony of simple, unconscious neurons that, when connected together in a certain way, function as a conscious whole. And yet, whenever we make an observation about intelligence in nature, we see that complexity is never enough. For interesting, adaptive behaviour to emerge, the system must also have the ability to self-organise. Self-organisation is what distinguishes cybernetic systems from other non-cybernetic complex systems. Self-organisation is not a result of complexity, but the other way round. Self-organisation occurs because the individual parts of a system connect in such a way as to create multiple levels of positive feedback. These loops then create ever-higher levels of self-organising complexity, and therefore new behaviours. Complexity of the interesting kind is the result of self-organisation based on simple rules. Such self-generating complexity sometimes gives rise to intelligent behaviour, and at least once in the evolutionary history of our planet has given rise to higher consciousness. But could such complexity also arise in interconnected computer systems?

This is a very hard question to answer, since we do not yet have a way to collect credible evidence.[26] Nevertheless, I personally would be inclined to bet that the spontaneous emergence of self-awareness in current technological cyberspace is highly improbable. Since the 1940s, we have been locked in a specific approach to computer technology that separates hardware from software, and which is mostly based on a specific hardware architecture called the 'von Neumann architecture', as we saw in the previous chapter. There could have been many other paths we could have taken in computer evolution (for instance advanced analogue computers), but we did not. The word 'evolution' is of great importance here. The pseudo-cybernetic assumption of the AI Singularity hypothesis essentially claims that an evolutionary kind of emergence of self-awareness is involved. Let us accept, for argument's sake, that evolutionary forces are at play in the way computers have evolved since the 1940s. Perhaps, as computers adapt to the changing circumstances of economic and social life of humans, they effectively 'evolve', albeit with the heavy intervention of their human designers. Nevertheless, even if we accept that tenuous proposition, evolution is not a one-way ticket to self-awareness. Out of all the millions of species that have evolved on this planet over the past four billion years very few seem to have achieved self-awareness, and only one is both self-aware and intelligent enough to build computers. However, that does not mean that species that are not self-aware are less 'complex' than humans. For all we know, the vast majority of species on Earth have evolved to be highly complex and not be self-aware. Indeed, there are non-living systems that are even more complex. Earth's ecosystem is more far more complex that all its individual parts – including us humans – but our planet does not possess self-awareness.[27] Computer systems will further evolve in ever more complex architectural and connectionist configurations, but this does not signal the teleological emergence of computer self-awareness in the future.

To conclude, both fundamental assumptions for the AI Singularity appear to be highly problematic. For AI to take over the world it must first become self-aware – or 'awake', to use Vinge's own term.[28] Nothing in the current technology points even remotely towards such an eventuality. Computers may be becoming increasingly more powerful in terms of calculations per second,[29] and able to perform tasks demanding increasingly intricate levels of knowledge, but they are still a long way from doing what a human baby can do without even thinking. When was the last time you saw a computer giggle at a funny face?

The Moravec paradox

The inability of computers to perform basic human functions has been succinctly defined by AI researcher Hans Moravec as a paradox. He writes: '. . . it is comparatively easy to make computers exhibit adult level performance on intelligence test or playing checkers, and difficult or impossible to give them the skills of a one-year-old when it comes to perception and mobility'.[30]

Although computational methods can reproduce high-level reasoning – as demonstrated in the case of expert systems – research in robotics has shown that sensorimotor skills remain a huge challenge. Coding cognition has proved to be an easy problem. The really hard problem in AI is coding sensing and action. According to cognitive psychologist Steven Pinker, this is the most significant discovery about AI.[31] It suggests that in the second machine age, while lawyers and doctors may struggle on social benefits, gardeners and janitors will remain in business and thrive. But why is this so?

Many AI researchers, including former MIT professor and current robotics entrepreneur Rodney Brooks, point out that human sensorimotor skills are not related to cognition but are the product of millions of years of evolution.[32] Despite the success achieved in AI by approaching the problem of intelligence from a different angle (the 'aeroplane' way), one would really need to reverse-engineer evolution in order to reproduce the full capabilities of a human brain including self-awareness and high-levels of consciousness. This realisation challenges the theoretical foundations of Artificial Intelligence, which are based on symbolic logic. Ever since the Dartmouth conference in 1956, AI researchers have assumed that the world can be represented with symbolic logic. We saw how the discovery that symbols can be used to construct logical operations led to the development of computer languages, and those languages were then used to represent knowledge about the world. Knowledge about the real world is *sine qua non* for furnishing computers with intelligence. Alas, the Moravec paradox illustrates the failure of this approach. The best that the most common home robot can hope to achieve today, after decades of robotics research, is to clean your floor – as long as you do not ask it to climb the stairs. This means that there are many types of knowledge that cannot be adequately represented with symbolic logic.

Sensorimotor skills are not the only problem area in AI, where symbolic logic seems inadequate to represent the world. Take, for instance, creativity. Humans are infinitely better than computers at visualising new ideas in science, art or everyday life. Even in chess, highly skilled human players using computers always win when playing against stand-alone computers, regardless how powerful their mechanical opponents might be. The reason humans always win is that they use their computer to explore deep moves, then use their creativity to strategise in unexpected ways. This clearly demonstrates that creativity goes beyond brute computing power. Maybe this is because human creativity requires us to accept conflicting ideas as being true at the same time. As F. Scott Fitzgerald famously stated 'the test of a first-rate intelligence is the ability to hold two opposed ideas in mind at the same time and still retain the ability to function'.[33] This is different from holding a number of possible uncertain hypotheses – a problem that can be coded for using Bayes theorem. Usually, we humans remain unconscious of the fact that we hold conflicting beliefs. Aristotelian logic, however, upon which computers function, dictates that if something is true its negation must be false. A computer cannot hold conflicting beliefs. In a computer program a statement can only have one truth value. Let's examine this 'paradox' a little more deeply, since it is very informative of the limits of quantifiable, symbolic, mathematical, computer logic. Let us think of what moral philosophers call 'hard choices'.[34]

Think of questions such as what career to follow, whether to break up or get married, move to the country or stay in the city, and so on. The problem with making such choices is that you cannot quantify their outcomes. That's what makes them hard. Choosing to study law instead of going to art school may make quantifiable sense if one compares the median incomes of lawyers to the median incomes of artists. But if everybody were to choose their career path that way, no one would ever study art. There is something unquantifiable that makes people decide to follow a profession that is likely to bring them much disappointment and a pittance by way of remuneration. Given a hard choice such as this, our decision can only be taken on the basis of what we stand for; of who we are; of what kind of person we aspire to be. These deep moral decisions require an 'I' that constantly strives towards a higher moral goal. This 'I', our 'self', holds a belief about who he or she is, and acts according to this belief. Indeed, hard choices are what make us who we are. Without them we drift opportunistically, bereft of a moral compass, regretfully achieving less than our full potential of being human. You cannot live a moral life using a mathematical formula. And that is because when something is unquantifiable it cannot be operated upon in any mathematical or logical way. Our moral decisions are therefore not logical. A purely logical being would have an issue understanding another being motivated by moral values – a fact amusingly illustrated by the comic dialogue Mr Spock and Dr McCoy often have in the original *Star Trek* series.

The Moravec paradox, creativity and hard choices demonstrate that the most essential aspects of being human remain beyond logical representation. This should give pause to anyone claiming that computers based on current technology will surpass human intelligence by 2030, or indeed ever. The American philosopher Hubert Dreyfus, one of the harshest critics of AI, claims that human intelligence mostly depends on unconscious instincts rather than conscious symbolic manipulation, and therefore cannot be captured in formal rules.[35] Modern neuroscience has vindicated Dreyfus. As discussed in Part II of this book, consciousness is not a logical algorithm executed by brain cells but the result of a multitude of chaotic and unconscious mental processes becoming integrated in the neocortex. By adopting the dualistic separation of software and hardware, computer science has managed to produce amazing machines that have shaped our world. At the same time, however, it departed from how an actual, unitary, biological person processes information. AI researchers have not only adopted the software/hardware paradigm, but have also adopted the brain-in-a-vat paradigm that disembodies intelligence and reduces it to processes taking place in a distinct system (the brain, the computer processor) only indirectly interfaced with the outside world. The biological reality is very different. Our brain is an integral part of our body that interfaces seamlessly with every other system in our body, including the circulatory, hormonal and the peripheral nervous systems. Sensation and action are processed and communicated using multiple distributed systems throughout our body. We are our body, and our consciousness is the integration of a corporeal experience in continuous interaction with our environment, our sensations and our selves.

And this leads us to an inevitable conclusion: that if we want to engineer a conscious machine we have already reached the limits of conventional computer technology. Symbolic representation can only take us so far. It does not matter how many quintillion calculations per second computers will be capable of performing by the next decade. Computer technologies that are based on separating hardware from software and which use symbolic logic to represent the world may become intelligent enough to replace many knowledge-based jobs, but they will never become conscious. They will therefore not threaten the survival of the human race, although they may affect our political system, as I will discuss at the end of the book. But if the best of our existing computer technologies are inadequate when it comes to the likelihood of their ever achieving sentience, are there some other technologies that might succeed where they fail?

To answer this question we must depart from the software–hardware duality and explore some alternative approaches to Artificial Intelligence. We must look back into how nature, the greatest engineer of all, did it. Starting with very simple chemical ingredients, nature first began to evolve conscious artefacts around four billion years ago. She has therefore shown us that it is feasible to create mind from matter. If we wish to mimic her we must reinvent Artificial Intelligence to follow her example. We must return to the basic principles of cybernetics and begin the journey anew, questioning our assumptions and doubting our apparent successes. If we want to develop true machine intelligence we must open the doors of the AI lab to life, evolution, and chaos.

16

DARWIN AT THE EDGE OF CHAOS

In our quest for alternatives towards developing sentient machines there are three areas where we must look for guidance. The first, and perhaps most obvious, is the human brain. In Part II we explored the nature and science of the mind, and saw how philosophers and neuroscientists have probed into the mystery of consciousness. And yet it seems telling that technology that aspires to emulate the human mind remains 'locked in' the wrong paradigm, perhaps because it has managed to achieve impressive practical results in so many other areas. The fact is that the computers we use today to perform 'intelligent' tasks bear no resemblance to the one thing in the world we know for sure to be truly intelligent: our brain. Nevertheless, scientists and engineers have been experimenting with computer architectures that mimic the central nervous system ever since the 1950s. I will examine the most promising of these attempts and reflect on what they may hold for the future of Artificial Intelligence.

The second area to explore is the mechanism by which life evolves and how it could relate to Artificial Intelligence. Perhaps we could begin with building extremely simple machines, apply the lessons of evolution and wait for natural selection to do the rest. Take, for instance, reproduction – sexual or asexual – a mechanism that drives biological evolution. Living organisms evolve by adapting their genetic make-up to changes in their environment through a process called mutation. Successful adaptations, or mutations, are then passed on to the next generation through reproduction. We saw how John von Neumann invented the Universal

Constructor, a self-replicating machine that can reproduce itself like a unicellular organism. It is therefore theoretically feasible to create the digital equivalent of a microorganism. Will this new, artificial life form evolve over time? Thankfully we do not need to wait for billions of years to find out. We can test the theory by massively accelerating evolution in a digital medium. Digital programs – imagine them as digital microorganisms – can access vast memory resources and data at very high speeds. They can search the computational space of all possible mutations and instantaneously find the mutation that is most beneficial for thriving in a changing environment, i.e. without having to undergo pointless trial-by-error experimentations over many generations. As we shall see, such experiments already take place not only in a simulated digital environment but also in the physical three-dimensional world. Today there exist nanorobots that can follow simple instructions and gradually assemble themselves into bigger artificial superorganisms that exhibit intelligent adaptive behaviour, or 'swarm intelligence'. Are these nanorobots the harbingers of a second genesis on Earth? Do we live at a time when a new kind of life is shooting out roots into our world, totally unsuspected? What might the future of this mechanical life form be? Could it evolve to become smarter than us? Could artificial life be a prerequisite for Artificial Intelligence?

Evolving computational machines with brain-like organisation presuppose a deep connection between computation and life. The human brain evolved, and so did consciousness, thanks to complex combinations of chemical elements. These combinations are essentially mathematical and algorithmic. Although, as we saw, symbolic logic falls short when it comes to representing the world meaningfully, nature follows logical rules and performs intricate calculations. This appears to be a contradiction. We cannot use symbolic logic to code for common sense. And yet logic and mathematics dictate the way bird flocks swirl and why bees build their hives as stacks of perfect hexagons – to give just two of the numerous examples of nature behaving like a computer. Once again we crash into the Platonic–Aristotelian dichotomy: is it form that takes precedence over matter, or vice versa? Does nature calculate compelled by the 'stuff' that our universe is made of, or is our universe the result of calculations? But what if this dichotomy, which has dominated Western thinking for centuries, is a false premise? What if there is no real distinction between matter and form, and it is the language we use to describe these notions that confuses us? No wonder the neopositivists of the Vienna Circle tried to outlaw language and adopt purely abstract symbols to describe the world. Unfortunately, their approach failed; and AI systems so far are a testament to this failure. Abstract symbolic reasoning cannot produce meaning on its own. The observed and the observer must somehow become one in order for meaning to emerge out of meaningless representations.

Nevertheless, new scientific research is shedding fresh light on to how biological organisms achieve homeostasis, the cybernetic state of dynamic equilibrium – which is essentially a synonym for life. The observed becomes one with the observer in homeostasis, as they become entwined in an eternal braid of self-referencing. What useful lessons can we learn from this new knowledge and apply to Artificial Intelligence? Could we construct homeostatic machines? And, if we do, shouldn't we accept them as 'alive', as the new members of Earth's biota?

Neural machines

Great insights in science often arise from observing similarities between seemingly uncorrelated phenomena. One of the most profound of such insights was garnered by the American biologist Gerald Edelman (1929–2014) who shared the Nobel Prize in 1972 with Rodney Porter, for discovering the structure of antibody molecules and explaining how the immune system functions. What Edelman discovered was that our body has a great number of structurally different antibody cells. When a bacterium or a virus enters our body these antibody cells[1] rush towards them and test how well their structures 'match' those of the intruders. Structural variability lies at the heart of antibody-based recognition. Edelman noticed that the adaptive immune response bore the hallmarks of an evolutionary process. The antibody recognition system 'evolved' quickly in order to adapt to the bacterial or viral attack. The numbers of the evolved antibodies also increased, since they were more successful in adapting to the attack. The immune system behaved like a species adapting to environmental pressure and evolving new functions or behaviours.

Edelman's research into antibodies led him to realise the enormous explanatory potential of selective-recognition systems. He posited that this evolutionary biological mechanism could also explain consciousness. Two significant discoveries strengthened his hypothesis. Firstly, that a fundamental property of cortical neurons is that they are organised into discrete groups of cells. Secondly, that synapses are strengthened by repeated use. This strengthening of 'successful' synaptic pathways takes place through feedback loops; the same information travels along the same path again and again, and this event is fed back repeatedly in order to keep that particular neural channel open. Edelman theorised that our brain manages to recognise and process information thanks to the selection of neuron groups that vary in their connectivity patterns. Several group cells would respond to incoming sensory information; their response would be modified by repetitive recognition that strengthened, abstracted and associated their connectivity. In effect, Edelman's theory of neuronal group selection described the brain as a cybernetic system with multiple positive feedback loops (although he always persisted calling them 're-entry' loops). As we saw in Part II, recent research by Stanislas Dehaene on the neural correlates – or 'signatures' – of consciousness illustrates that re-entry mechanisms are fundamental to how groups of cells respond to sensory information, and explains how a local recognition event becomes global by spreading throughout the cortex and entering our awareness.

To demonstrate his theory, Edelman and his colleagues at the Institute of Neurosciences at La Jolla, California, built a number of 'noetic machines' called 'Darwins', or 'brain-based devices' (BBDs). Modelled on the neural connectivity of a simple brain, a Darwin robot 'discovers' the world around it in the way that an animal would. The essential part of a BBD is a neural simulation that takes place either on its on-board computer or, via Wi-Fi, on a remote and more powerful computer. This digital neural simulation, unlike a conventional software program, is capable of learning through modifications of simulated synaptic strengths based on the outcomes of previous actions. A value system ensures that a selection takes place, just like it does in the brain. For instance, it is 'better' (higher value) to have an object in the visual field of a robot than having a uniform visual field (lower value). These value systems are quite general. When the robot does something 'good' in terms of value, the synaptic strength changes so that, in a similar situation next time, the same 'good' action is more likely to happen again. By applying this simple principle the robots can perform quite complex actions. The change in synaptic activity is known as 'neural plasticity' amongst neuroscientists and neurologists, and is an essential property of the brain. We learn by constantly modifying the relative strengths across our neuronal groups. When someone suffers a brain trauma, it is often the case that neurons will rewire themselves in an attempt to compensate for functional loss, thanks to the brain's plasticity. It has been shown that brain plasticity is something that we can consciously affect, at least to a certain degree. When we learn how to play a musical instrument, win at chess, or perform mental calculations, we are forcing our neurons towards a state that facilitates the better performance of these tasks.

Although Edelman's robots aim to explore the theory of neuronal group selection, they represent an interesting case of robots interacting with the natural environment, and learning, by simulating the brain. These 'neural robots' are more about science than engineering, and that is perhaps the reason why Edelman's machines have not gained much traction in the wider robotics community. Nevertheless, mimicking the brain has been a core theme and obsession in Artificial Intelligence ever since its infancy. The connectionist architectures of the 1960s and 1970s aspired to develop computers based on artificial neurons. The driving principle behind those prototype artificial neural networks was not dissimilar to Edelman's approach. The difference was that these artificial neural networks were hardware configurations that received an input, applied an algorithm that evaluated and strengthened the input information, and which then passed the result on to the next level of the artificial neural network. The process was repeated several times until the end result was to correctly recognise the input. To achieve this result the network had to be 'trained', which in effect entailed providing feedback to the network about how closely its pattern matching was to the original input. This feedback was then exploited by the evaluating algorithms to simulate neural plasticity. Several repetitions later, the artificial neural network has 'learned' how to perform a good match between input and output – in other words it has learned how to correctly recognise the input.

Artificial neural networks have since been used to perform pattern recognition in visual systems, machine learning, as well as other applications in which it has been difficult to code in a conventional way. But the approach was more or less abandoned by the late 1990s as new and more sophisticated statistical and signal-processing techniques arrived on the scene, which could perform most pattern recognition tasks satisfactorily using conventional computer architectures.

However, connectionism has enjoyed a spectacular comeback in recent years. An important innovation in electronics called 'memristor' has played a key role in the revival of neural computing. Originally, the memristor was described in a 1971 paper by electronics theorist Leon Chua[2] as a variable resistor that 'remembers' its last value when the power supply is turned off. Although Chua argued that the memristor was a fundamental circuit element (like a capacitor, an inductor or a resistor), its physical realisation eluded engineers for decades. Until in March 2012, that is, when a team of researchers from HLR Laboratories and the University of Michigan announced[3] the first functioning memristor array built on conventional chips.[4]

Connect a memristor with a capacitor and you get a 'neuristor',[5] an electronics component that behaves like a neuron. This is how a neuristor works: as an electric current passes through the memristor its resistance increases and it heats up. At the same time, that capacitor that is coupled to the memristor becomes charged. However, the memristor has an interesting property: at a given current threshold its resistance suddenly drops off, which causes the capacitor to discharge. This discharge – or 'spike' – at a certain threshold is akin to a single neuron firing. Memristors and neuristors are elementary circuit elements that could be used to build a new generation of computers that mimic the brain, the so-called 'neuromorphic computers'. These computers will differ from conventional architectures in a significant way. They will be essentially analogue rather than digital, thus eliminating the present software–hardware dichotomy. Such analogue circuits with their ability to 'spike' will mimic the neurobiological architecture of the brain by exchanging spikes instead of bits. There are various technical challenges in advancing memristors and neuristors to the level at which they could be used to construct sophisticated, and practically useful, computational machines. Although we know how to integrate information using software algorithms and symbolic logic, research is still being carried out into how we could integrate instances of spikes across vast arrays of neuristors. Neuromorphic technology is still in its infancy; however, its future development is of great international interest and a main research goal of both the European Human Brain Project and the American BRAIN project.

There are several good reasons for this transatlantic interest. It is estimated that in order to simulate the human brain on a conventional computer we will need supercomputers a thousand times more powerful that the ones we have today. This requirement is stretching the limits of the current technology in chip manufacturing, and for many it lies beyond the upper forecast of Moore's Law. But memristors and neuromorphic circuits could push computing's envelope a lot further. The experimental memrisitor built by the HLR Laboratories and Michigan University group demonstrated memory storage of thirty gigabytes per square centimetre, which is three thousand times greater compared with the ten megabytes per square centimetre of current rigid memory disks. This measurement alone indicates the huge potential of these new electronics components.

If neuromorphic architectures prove to be the future of computing then the supercomputers of the mid-twenty-first century will be more brain-like, and computing will transcend the current division of hardware and software. The software will be the hardware. Digital processing will become a thing of the past, as the computers of the future will be capable of learning and responding to their environment by utilising their built-in plasticity, similarly to biological neural networks. But will these neural computers be able to evolve and become conscious?

The Invincible

In the summer of 2014, Michael Rubenstein and his co-workers at Harvard University programmed 1,024 robots[6] with a simple set of instructions that compelled them to collectively form any shape given to them.[7] The 'kilobots', as they were called, simulated how flocks of birds or schools of fish behave. Four 'seed robots' acted as a point of origin for a coordinate system. Their coordinates were communicated to the kilobots nearest to them using infrared light. By this simple algorithm the information spread through the whole group and every kilobot adjusted its position by calculating its relative location in the swarm. It took approximately twelve hours for the swarm to form a shape such as a letter, or a star. Rubenstein's robot swarm is the largest yet to demonstrate collective adaptive behaviour and represents one step forward in the direction of artificial swarm intelligence. Once again, literature imagined this technology several decades ago.

In his novel *The Invincible* (1964), the Polish writer Stanisław Lem (1921–2006) tells the story of a spaceship that lands on a distant planet called Regis III with the aim of finding a mechanical life form, the product of millions of years of mechanical evolution. Lem's mechanical creature exhibits swarm intelligence: relatively 'dumb' parts self-organise into a superorganism with superintelligence. The parts are self-replicating, insect-like nanomachines which have survived a protracted robot war. As the ultimate victors, they had become the robotic fauna of the planet, responding aggressively and spontaneously to any threat. Lem was keen to argue in his novel that evolution does not necessarily lead to a species with superior intellect. Human evolution can only be regarded as accidental. Perhaps on another planet, like Regis III, nanorobots have taken over from their creators? Perhaps Regis III is Earth in the future?

'Nouvelle AI', a concept championed by robotics pioneer Rodney Brooks, argues that instead of trying to reproduce human intelligence in AI, we should focus on creating robots possessed of an insect-like intelligence that is capable of evolving.[8] Nanotechnologists, like the visionary Eric Drexler,[9] see the future of intelligent machines at the level of molecules:

tiny robots that evolve and – as happens in Lem's novel – which come together to form intelligent superorganisms. But in order for a life form to evolve it must reproduce or, in the case of machines, to have the ability of sexual, or asexual, self-replication.

Robot sex

The concept of self-replicating machines is neither modern nor pertains only to a distant, scientifically fictional future. The possibility of such a device was indirectly proposed in 1802 when English theologian William Paley (1743–1805) formulated the first teleological argument that touched upon machines producing other machines. In his book *Evidence of the Existence and Attributes of the Deity* he put forward the notorious 'watchmaker analogy' that has been quoted by diehard creationists ever since. He argued that something as complex as a watch could only exist if there were a watchmaker. For Paley, since the universe and all living beings were far more complex than a watch, it followed that there had to be an Intelligent Designer, a Divine Watchmaker, a God. Interestingly, Paley conceded that his argument would be moot if the watch could make itself. This small but significant detail in Paley's argument was all but lost during the cultural wars that followed Darwin's publication of *On the Origin of Species* in 1859. Evolutionary theory suggests the complexity of life can result without the need for design. It has shown how complex intelligent beings can evolve without the need for a Creator. Nevertheless, given that complex non-living systems can – theoretically – reproduce, Paley's subtle point about design appears less crazy that most evolutionists would like. For, if watches could reproduce, then watches could also evolve.[10]

The Victorian novelist Samuel Butler (1835–1902) took Paley's argument to another level. A contemporary of Charles Darwin, he spent twenty years of his life attacking the foundations of Darwinism. Butler was not so much against the idea of evolution per se. His tiff with Darwin revolved around the role of intelligence. For Butler, intelligence and evolution shared common principles since life was at the same time both the cause and the result. Four years after the publication of *On the Origin of Species*, Butler published an article in the New Zealand newspaper *The Press* entitled 'Darwin Among the Machines'.[11] In the article he concluded: 'It was the race of the intelligent machines and not the race of men which would be the next step in evolution . . . . In the course of ages we shall find ourselves the inferior race.' Butler's notions presaged the current debate on the AI Singularity, and have informed it in several ways. In his novel *Erewhon* (an anagram of 'nowhere'), he describes a utopian society that has opted to banish machines, which are deemed to endanger the survival of the human race. As part of the backstory to his epic novel *Dune*, Frank Herbert coined the term 'Butlerian jihad' to describe how thinking machines were outlawed 10,000 years before the book's main events. Both Butler and Herbert suggest that the only way we can avoid being replaced by our creations is to oust them before it's too late. The narrative of fear of the intelligent artefact, exploited by talented writers such as Butler and Herbert, becomes even more potent when scientific advances appear to confirm the predictions of science fiction.

John von Neumann's Universal Constructor is key to understanding how artificial life could reproduce. The Universal Constructor is both an active component of the construction as well as the target of the copying process. The medium of replication is stored in the instructions for the replication. This allows open-ended complexity and therefore permits errors in the replication, thus opening up self-replicating non-biological systems to the laws of evolution. Mechanical engineers have taken von Neumann's principle on board and have produced physical applications such as RepRap machines,[12] which are 3D printers that can print most of their own components. There is another way of robots reproducing themselves, too.

Imagine a robotic factory with three classes of robots: one for mining and transporting raw material, one for assembling raw materials into finished robots and one for designing processes and products. The latter class, the 'brains' of the autonomous robotic factory, would consist of Artificial Intelligence systems. Such a robotic factory would have the ability to evolve by design; the 'artificial brains' would continuously optimise the worker robots on the basis of some value system, for instance to extract the maximum amount of minerals for the least expense of energy. Currently, existing safety legislation in most developed countries impedes, although it does not preclude, the development of a fully autonomous robotic factory capable of reproducing itself. But planting such a factory on a distant planet is a different story. The colonisation of Mars, for instance, could benefit from self-reproducing robots designed to prepare the planet for human habitation. Back in the 1970s, Freeman Dyson proposed using self-replicating robots in order to cut and ferry ice from Engeladus (a frozen Saturn satellite) to Mars, using the ice to terraform it.[13]

Are von Neumann's Universal Constructor and Rubenstein's swarming kilobots signs of the imminent realisation of one more sci-fi prophecy? Or are they crude approximations, hopeless attempts, to achieve something that intelligent design cannot possibly reproduce: artificial life. If they are the latter we have nothing to fear. We can indulge ourselves like children with stories about terrifying robots, and scare each other for laughs. We can use kilobots as toys. But if it is possible that simple self-replicating machines could evolve then we will have to rethink how to proceed from here on. Although we have nothing to fear from conventional computer technologies, neuromorphic computers and nanorobots usher in a completely new set of circumstances. By continuing to pursue the development of these technologies we may indeed be sowing the seeds of a new evolutionary force that will ultimately exclude us. To answer this very vital question we need to abstract the concept of life.

What is life? What separates the living from the non-living? What mechanism makes a complex system of chemicals counter the natural law of entropy and exchange heat with the environment so that it keeps its internal organisation intact? To discover this mechanism we must re-enter the world of cybernetics and journey to the edge of chaos, the place where life and intelligence meet – just like Butler predicted.

The algorithms of life

Life is replete with self-organisation phenomena whereby fundamental building blocks form intricate networks that exhibit new behaviours and functions. Amino acids determine the structure of proteins, genes the fate of cells and neurons think and write books. What all these phenomena have in common is that the parts come together in a very special state called 'dynamic equilibrium'. Like madmen, the fundamental building blocks of life dance together at the edge of an abyss. From the outside, it looks as if one small nudge could end the dance abruptly, making the parts scatter and the whole disintegrate. But it doesn't. In fact, once the parts reach this critical state their collective behaviour becomes highly susceptible to external perturbations and at the same time highly resilient. If you have ever watched a flock of starlings responding to an attacking predator you have surely marvelled at how each bird knows to move swiftly and in unison with its peers, never abandoning the flock. As if caught up in their collective celestial dance, starlings faithfully keep to their ranks. Delve deeper and you will discover that each bird – just like each one of us humans – is alive because its body is also made up of fundamental building blocks that come together in dynamic equilibrium, a state called homeostasis. What happens at the level of a single cell, or a single bird, happens at the level of the flock as well.

Biology involves many of these critical networks forming multiple hierarchies. Like Russian dolls stacked one inside the other, our existence is predicated by a succession of dynamic systems, all of which walk a tightrope separating order from chaos, with death lurking but one false step away.[14] But, why does life need to exist at the edge of chaos? What compels the parts to form hierarchical networks? How do they achieve this very fine and precarious state of criticality?

We are still ignorant when it comes to the answers to these very important questions about life. However, phenomena of criticality exist in other areas of scientific enquiry as well, for example in the spontaneous magnetisation phenomena, or when crystals form. These phenomena have been studied for years using a host of scientific and mathematical techniques. What makes the study of similar phenomena in biology exceptionally challenging is lack of verifiable experimental data. Biological systems are the most complex in the universe, and extremely hard to isolate. Everything is connected to everything else, most often in ways we have not yet imagined. The fact that biological systems are poised in a state of criticality begs the question of whether there is a universal law behind all this. If such law exists, and we could find it, then we would have discovered how to create bodies, minds and thoughts. Artificial Intelligence would be straightforwardly reduced to programming neuromorphic computers with this ur-algorithm.

Curiosity about the existence of a general law in biology dates back to the Macy Conferences. In an influential paper published in 1959, Warren McCulloch, Walter Pitts and Jerry Lettvin explored the visual system of the frog.[15] They showed that the eyes of the frog do not actually represent reality but construct it. And what is true for frogs must also be true for humans. Our eyes and nervous system construct our idea of reality. This idea is constantly checked and verified as sensory signals feed back to our nervous system. The observer and the observed are therefore one: we are the reality we observe, because our internal biological structure is this observed reality. The fourth co-author of that paper, a young neurophysiologist from Chile called Humberto Maturana, expanded on this idea to develop a cybernetic concept for cognition called 'autopoiesis'.[16] He, together with his student Francisco Varela, coined the term autopoiesis in 1972 to denote living systems that maintain their internal structure while in constant interaction with their environment.[17] Maturana and Varela offered the example of a typical cell, made of various components such as nucleic acids and proteins. These components form intricate structures: a nucleus, a cytoskeleton, a membrane. What Maturana and Varela showed was that the structures created the parts and the parts created the structures; that the cell constantly made itself from itself, by itself. This mechanism contrasts with what 'allopoietic systems' do. They use raw materials to construct something else, but not themselves. A typical example of an allopoietic system is a car factory: it makes cars but the cars do not make the factory.

Maturana and Varela also showed how autopoiesis takes place through the external flow of molecules and energy. Generalising on their insight, they suggested that all living systems are autopoietic: their goal is to exchange energy with the environment in order to sustain their internal structures. Equilibrium between internal states and external environment occurs at criticality. Autopoiesis defines death as the breaking of criticality. When this happens there is a spontaneous separation of subject from object, of the observer from the observed. We die when we stop constructing reality, which is the same as saying when we stop constructing ourselves. So is autopoiesis the general law of life?

Autopoiesis has received considerable criticism, mostly because it claims to solve a central problem in epistemology by making the observer part of the observed. It does not, however, answer the question as to how it all started.[18] In the short story 'Circular Ruins', Jorge Luis Borges gives us a literary illustration of autopoiesis, as well as its weakness. A wizard, a member of a long line of ancient wizards, clones himself using magical rites. To do so he is prevented from remembering his past. The wizard is a total stranger, including unto himself. He arrives at a magical temple by following his instincts, like a salmon returning to its breeding ground. Using clay, he constructs his clone in a half-sleep state, reciting incantations to the gods, speaking magical words. The clone is created, an imperfect Adam, and the wizard trains him and then sets him off into the world. In the years that follow the wizard hears of his protégé's successes in the world, and realises that his mission is accomplished. He returns to the temple and self-immolates. One day his clone will return here to do exactly the same thing, the circle repeating forevermore. Autopoiesis looks like Borges's circle of wizards being born and dying in the temple. But how was the first wizard made?

Another student of the legendary Walter McCulloch claims to have an answer to that question. Stuart Kauffman, currently at the University of Vermont, studied a class of chemical reactions in which the product of reaction is the catalyst of the reaction. These so-called 'autocatalytic' reactions may explain the origins of life – or how the first autopoietic wizard in Borges's story came about.[19] Experiments have shown[20] how, through autocatalysis, chemical ingredients can exhibit a rudimentary form of natural selection. Kauffman's research on the mathematics of autocatalytic reactions has shown that these reactions take place at criticality. The mathematics of these reactions involves recursive computations emerging spontaneously at the edge of chaos. It looks as if nature is a lover of extreme sports. It adores pushing everything that is precious to the point of breaking up. This deeper link between the emergence of complex behaviour at criticality and recursive computations has also been demonstrated in digital computers using cellular automata, another great invention by John von Neumann. Cellular automata are patterns of 0s and 1s that evolve step-by-step according to a simple set of rules. A new pattern, or 'generation', of a cellular automaton emerges after each step. Points on the new patterns will be either 0 or 1 depending on their current value as well as the value of their neighbours.

In the early 1980s, the English mathematician Stephen Wolfram conjectured that a particular cellular automaton called 'Rule 110' might be 'Turing complete',[21] a conjecture that was later proved by Matthew Cook. 'Turing complete' means that Rule 110 is capable of universal computation, i.e. any calculation or computer program can be simulated using this automaton. What is particularly interesting about Rule 110 is its behaviour on the boundary between stability and chaos. It is neither stable nor completely chaotic. Localised structures appear and interact in various complicated looking ways. Its behaviour looks so cunningly lifelike that it has led Wolfram and others into reverse thinking, and consequently to the conclusion that life is a program run on a cosmic digital computer. Platonic thinking, so deeply entrenched in mathematical culture, is evidently at play here! Nevertheless, the correlation between computation and life is indisputable. It is the interpretation of this connection that polarises empiricists and idealists. Perhaps the discovery of Rule 110 is one giant step towards the discovery of a general, mathematical, law for life. There are too many things about cellular automata that make them profoundly similar to physical, living, things. By operating near the edge of chaos, cellular automata evolve with time by responding to their changing environment. They look like a form of 'artificial life' existing in the computer that runs the calculations that make and sustain it. Could this artificial life evolve to the point of becoming conscious? And, if so, how similar would this artificial consciousness be to ours?

Imagining true AI

Let us summarise what we have explored so far. Biological systems function at the edge of chaos, a mathematical boundary that separates them from self-collapse.[22] Biological systems are essentially autopoietic: their parts make their structures that make their parts. Research into autocatalytic reactions and cellular automata has shown how autopoietic systems might have emerged: given the right initial conditions simple local interactions become emergent features of extended hierarchical networks. In the case of autocatalytic reactions the right initial conditions would be simply having the ingredients in place within certain boundaries for temperature and pressure. The journey of the individual parts towards forming a self-organised system appears then to be algorithmically determined: they are 'attracted' to self-organisation and, ultimately, to life. We do not yet know whether this attraction is governed by a general law for biology. However, we have discovered something that seems to point towards such a law: Rule 110, a recursive algorithm that is Turing complete and lifelike – and there might be more.[23] This profound correlation between cellular automata and biological phenomena suggests that life is governed by recursive computations, probably similar – or identical – to cellular automata.

There is one more special feature of complex computations that is worth noting. They are fractal-like and scale-invariant. This means that they repeat themselves at every scale. From microscopic organisms to weather systems and the formation of galactic clusters nature creates similar patterns of organisation and behaviour. These patterns transcend the boundaries we place in order to separate the living from the non-living. Just think of our circulatory system and how our veins bifurcate in our body, and then observe exactly the same pattern in the deltas of rivers. Both patterns are the result of similar complex computational iterations at the level of cells in the case of our body and of water molecules in the case of the rivers. Future developments in neuromorphic computers may provide new opportunities for experimenting with such complex computations in order to simulate life and cognition. For instance, instead of writing software to represent knowledge about the world, we could perhaps compel interconnected neuristors to become autopoietic using cellular automata. Knowledge from the study of the neural correlates, or signatures, of consciousness will also be applied to this newly engineered, brain-like medium. Instead of simulating a human brain in a computer we will be building a physical replica from the bottom up, using hardware. The neuristors of the future may not be imprinted on solid-state materials, as they are today. Perhaps new technologies in liquid electronics or quantum computers will allow us to use materials that can float in a solution. These 'liquid' neuristors could then be 'pushed' by complex computations to the edge of chaos, and acquire capabilities similar to interconnected biological neurons. They would then be able to form hierarchical complex neural structures, replete with plasticity and the ability to self-organise, culminating in a fully functioning artificial brain.

This brain would then have to be somehow embodied. Like any living thing, it will have to sense its environment in order to learn and develop. Depending on how far the miniaturisation of liquid electronics goes in the future, these artificial brains could be installed inside biomechanical bodies capable of homeostasis, and thus we will have arrived at a time when androids will become a reality. Alternatively, artificial brains could exist, at least in the beginning, inside nourishment machines, like those philosophical brains-in-a-vat we have discussed in Part III. They could potentially interface with the outside world through distributed sensory networks. They would be able see, hear, smell and touch, as well as access vast databases and other computer resources, including communication with other similar artificial brains. These artificial brains of the future, whether they are androids or placed inside vats, will be the result of engineering design as much as of natural evolution. Their evolution will take place gradually. It will be very slow at first, but exponentially rapid later.

At first, these artificial brains would be very simple and rudimentary, perhaps similar to an insect's. But they would represent a major breakthrough in artificial life and intelligence. They would have demonstrated our ability to transfer the lessons we've learned from nature – about life, complex computations and cognition – to an artificial medium. The algorithms of life would run in those tiny mechanical brains as they huddled together, exploring the controlled environment of the AI lab. But, as we saw, the algorithms of life can be scaled up by following a power law: every new generation will be many times more evolved than the previous one. From having the intelligence of tiny insects, artificial brains will quickly acquire the intelligence of reptiles, birds, mammals, primates, and finally that of human beings. At that point we will have created a mechanical and intelligent creature in our own image.

In the beginning, this creature might be similar to a human newborn. It will need nourishment and care. Its human creators, or foster parents, will need to stimulate it with talk and play, so that it begins to learn our language and become social. Like human babies, those first artificial human-like minds will be unique, each and every one of them. And, unlike today's computer technologies that can copy themselves infinite times, the conscious AI of the future will not be able to replicate itself exactly. The internal structure of their artificial brain would be subject to the autopoietic laws that govern the structure of biological brains. Therefore, its structure will not be static but plastic, always adapting to external and internal stimuli, constantly rewiring itself so it remains functioning at the edge of chaos. It will, however, be able to create other artificial minds that are similar. It will also be capable of expanding itself and augmenting its capabilities. By absorbing chemical nutrients from the solution in which its electronics are immersed, it will constantly build new neuristors and other parts, new connections, new channels and new networks. Given the human value system that considers high intelligence to be a good thing, the artificial mind will strive to become ever more intelligent.

The machine psychologists of the future will no doubt argue about the best way to help those early artificial minds develop in order to reach their full potential. Some may be in favour of arduous training, and the infusion of human moral and social values that will guarantee the harmonious integration of those intelligent machines into human society. More radical machine psychologists might prefer to grant total freedom to the machines so they may discover their true identity. The stakes would be high. For, just like human beings, those machines will be in possession of free will. True Artificial Intelligence will not be based on symbolic logic and will therefore not be a formal logical system. These intelligent machines will be able to hold contradicting notions simultaneously, just as humans can. Gödel's incompleteness theorem will not apply to them. They will have an 'I'. They will probably enjoy mentally playing with logical paradoxes, like we do, perhaps because the edge of chaos that would define them, as it defines us, is a logical paradox. When they become adults, the machines will have to decide who they want to be, and what they want to stand for. Some machines will like our world and our values; but others might reject them. Others still may be completely indifferent to us. Their neural hardware will refuse to be shaped according to our wishes. They will ignore their teachers, be oblivious to the feelings of the human others, become confused with social interaction, and single-mindedly pursue their own interests. This latter category of intelligent machines may in fact prove to be the majority, if not the rule. They will exhibit reclusive types of behaviour similar to some of those we presently associate with autism.[24]

These behaviours could be the result of social selection. In his book *Average Is Over*, Tyler Cowen argues that success in the twenty-first century is already becoming synonymous with the ability to work with computers. My personal experience within the IT industry and Internet start-ups has been that the most brilliant programmers are often very shy of other people and social interaction. Science and technology have become so competitive nowadays that success in these professions most often comes to those exhibiting high IQs and maximum dedication to the lab or the computer terminal. For them, the bustling noise of crowds or pointless chitchat by the water cooler seem distractions from a rewarding career. People with some degree of what is called Asperger syndrome can do exceptionally well in a world where the top jobs go to the top programmers and top scientists. In fact, this is already happening, and may explain to a certain degree false perceptions about an 'autism epidemic'.[25] If this trend in social selection persists, the intelligent computers of the future will most likely be the foster children of families composed of highly intelligent autistics. This social selection process will be reinforced further by the very nature of cognitive autopoiesis, which produces closed systems for which the environment acts merely as a trigger for internal processes.[26] British psychologist Simon Baron-Cohen has hypothesised that autism correlates with an absence of theory of mind.[27] Viewed from a cybernetic point of view, this suggests that there are many alternative 'attractors' for cognition in the brain. Not all types of cognition arrive at the same point, where we get theory of mind. Most of us have theory of mind because of the selection processes that we saw in Part I of the book, when I discussed how our species evolved general-purpose language and general intelligence. However, throughout our long journey as a species there have been many of us whose cognitive systems achieved dynamic equilibrium at points where theory of mind was either absent or wanting. We identify this state of alternative cognitive equilibriums using the word 'autism'. But when it comes to intelligent machines evolution will favour 'autism'. It is therefore highly probable that the intelligent machines of the future will not have a theory of mind, and will therefore tend to ignore the intelligent primates who created them.

This does not necessarily mean that they will seek to exterminate us. But it is important to realise that whatever these intelligent machines do, or think, is simply unpredictable. Like us – autistic or not – they will exist on the edge of chaos. Some may decide to remain children and play. Others may become scientists and solve all of humanity's problems. Others may push the boundaries of the dynamic equilibrium that sustains them, fall in the abyss of self-destruction and re-emerge from it several orders of magnitude more intelligent. Let's remember that self-organisation phenomena scale according to a power law. The next stage in the evolution of intelligent machines is impossible to describe, imagine, or comprehend – because it will be many orders of magnitude higher than our intelligence. The distance between the intelligence of those machines and ours will be similar to what separates us from the ants. The much-dreaded AI Singularity will have arrived then, and all bets will be off.

We could put an end to all AI research that could lead to sentient machines even now. This could be effectuated via a global, collective decision sealed with an international treaty that banned Artificial Intelligence research – or to put it in more precise terms research in 'Artificial Consciousness' – forever. This would virtually obliterate the possibility of our world, and our species, disappearing because of an AI Singularity. An international treaty such as this is a very possible scenario in the decades to come. Until today, very few have taken seriously the threat of an AI Singularity. The arguments that I have explored in this book – such as that our conventional computer technology is highly unlikely ever to become sentient – influence the lack of political action. But one can easily imagine a time in the near future when a major breakthrough in neuromorphic computing raises the alarm. If one day we manage to create the first artificial conscious life forms, albeit of the very simple, insect-like kind that I alluded to earlier in this chapter, it will become profoundly obvious to everyone that we will have crossed a significant threshold in technology, one that could have devastating repercussions, a breakthrough in computer science similar to the splitting of the atom in the 1930s.[28] Such a major breakthrough might trigger a global debate that could lead to the banning of any further research. And that would be the end of our ancient dream to create living, conscious artefacts in our own image .... Unless, of course, by the time truly conscious machines come to the fore our world has become so different from what it is today that our reaction will be one not of alarm, but one of excitement and joy. Unless Artificial Intelligence, of the non-conscious type, will affect our culture and our political institutions to such a degree in the years to come that our value systems will change. In such a future scenario the emergence of conscious AI will be regarded not with horror but with eager anticipation for the greater promise it holds. Humanity thus transformed by Artificial Intelligence will make the AI Singularity inevitable. Let's now examine this future scenario in more detail, how it may come about and why it is more probable than we might have liked.

Epilogue: THE FUTURE OF HUMANITY

Artificial Intelligence is a technology unlike any other. Not only because of its potential to transform radically our societies, economies and the very planet that we live on; but because it is all about us – about who we are, how we think and communicate and what makes us human. I have argued that humanity's journey towards creating Artificial Intelligence started around 40,000 years ago in the painted caves of Europe and elsewhere.[1] Art was what the modern mind produced when it was born. The spark of artistic creativity ignited spontaneously because that's how our minds make sense of their own existence, as well as of others like us, and of the world. Artefacts that represented creatures of the imagination – ivory figurines such as the lion-man of Hohlenstein Stadel – were very much alive as far as our prehistoric forefathers were concerned. Such artistic creations were thought to possess minds and intentions of their own, and were significant nodes in an expanded social network that encompassed everything and everyone: humans, animals, plants, rocks, trees, stars, watercourses, artefacts – everything. Sculptures such as the lion-man were the prehistoric precursors of automata and robots. They were also the inevitable product of the four aspects that have characterised the modern mind ever since: dualism, theory of mind, story-telling and anthropomorphism. These aspects still define who we are today. They guide us as we develop technologies to improve our lives and gain more knowledge. Ultimately, these are the aspects that define our quest to understand ourselves, the greatest mystery of all. And what better way to explore this mystery than creating a thing that looks like us, behaves like us, speaks like us, and 'feels' like us: an artificial creature with intelligence made in our own image.

We have been talking about such imaginary creatures for thousands of years. Stories shared under the starry skies of prehistoric times passed from one generation to the next have become the myths we share today. Our novels and films retell these stories using new forms, characters, plots and contexts. Our brains are hard-wired for telling, and for listening to, stories. That is why narratives are the most powerful means available to our species for sharing values and knowledge across time and space. I have argued that narratives ultimately dictate our artistic and scientific endeavours throughout history. But this relationship is not one-directional. As we explore the world and discover new things, our narratives evolve as well. Our discoveries feed into our stories, and vice versa. One cannot separate the two. We strive to create Artificial Intelligence because we have been telling stories to each other about artificial beings ever since the Ice Age. For the ancient Greeks, these were about the gods who breathed life into inanimate matter. For the Romantics, it was electricity harvested from ferocious lightning during a stormy night. For the software engineers of the twenty-first century, the ghost in the machine is code. With time, the lion-man has been transformed into a cyborg.

Philosophy arrived late in our journey towards understanding how to build a mind. Mind philosophy arguably begins with Descartes in the seventeenth century. But mind philosophy's roots go much deeper, into the works of Plato and Aristotle, and to the dichotomy that those two ancient thinkers have bequeathed to our ways of thinking and culture. Because of them we are still unable to decide what takes precedence: form or matter? This dichotomy is further compounded by dualism, the idea that the world is essentially made up of two classes of things, one material and the other immaterial. Dualism remains popular with many empirical scientists today. Amazingly, science often seems to confirm that mathematics, i.e. something immaterial, rules the material universe. Our fundamental scientific understanding of the universe today rests on a purely mathematical theory whereby several geometrical shapes called 'strings' are thought to create matter and energy by twisting and twirling. Could the mind be something immaterial, too? The dualist idea that the mind is separate from the body has influenced modern computer architectures and the idea of information. The separation of hardware from software and the disembodiment of information are the direct results of the influence of philosophical dualism. As a consequence, we, too, have become binary creatures: post-humans living in the world of atoms as well as in a digital world of bits. There are many who seriously contemplate the possibility of downloading consciousness on to a computer and thereby achieving digital immortality. For many today, heaven has become a server in the cloud and the Christian soul has been replaced by software. The combination of ancient narratives and our inherent dualistic cognitive make-up is profoundly overwhelming – and constantly mutating.

Lessons from neuroscience and cybernetics

Nevertheless, empirical neuroscience reveals a very different picture about the mind. Mental processes in the brain create the mind as information from various anatomical areas becomes integrated in the neocortex. This integration happens because of multiple, hierarchical feedback loops that operate at very finite equilibrium points. What we call 'mind' is something that happens when mental process loops integrate information at the edge of chaos. Second-order cybernetics shows how one could simulate such events in a computer. But those events are not enough for high-level consciousness to emerge. For a mind to have a self it needs a singular body. Neuroscientist Antonio Damasio has argued that the feeling of self is the result of a permanently maintained physiological bond between the brain and the body.[2] He has highlighted the role played by the brain stem, a region embedded between the cerebral cortex and the spinal cord, which houses all the major life-regulation devices in the body. There are two different regions inside the brain stem which, when affected (say, by a stroke), result in completely different outcomes. One of them results in coma, i.e. the loss of consciousness. This happens when the body and the brain become totally disconnected. The other region, when affected, results in 'locked-in syndrome'; the terrifying condition whereby a patient retains consciousness but cannot move a muscle. This happens because although the control paths to the body have been severed, the brain remains 'aware' of bodily functions – and thus consciousness is preserved. Neuroscience tells us that a disembodied mind cannot exist – i.e. that dualism is nonsense. Consequently, a necessary (but not sufficient) condition for artificial consciousness is for an intelligent machine to have a unitary body.

But how close are we to creating such a machine? In 1978, as AI was entering its so called 'winter years', John McCarthy, the pioneer computer scientist who coined the term 'Artificial Intelligence', mused somewhat bitterly: '. . . human level AI might require 1.7 Einsteins, 2 Maxwells, 5 Faradays, and 0.3 Manhattan Projects'. Many AI researchers share this belief – that achieving AI is a matter of bringing enough talent and investment to bear on the subject. And I agree. Human-level AI *is* theoretically feasible, providing we explore new directions in computing that mimic brain function and take into account lessons learned from cybernetics and neuroscience. Neuromorphic computers might be the right direction to go, although these are still at a very early stage. What is definitely *not* the solution is to hope that existing computer technologies, because they double in power every eighteen months according to Moore's Law, will somehow evolve a mind, or a self, or consciousness spontaneously. They will not, and in this book I have explored the reasons why. Formal languages, such as the ones used by software engineers to program computers, are neither sufficient for meaning to emerge, nor for the world to be adequately represented in a machine. But if this is so, why do so many bright people today insist that human-level AI is just around the corner?

The simple answer is to be found in language, or, better still, in the evolutionary history of language. Our primate ancestors evolved specific-purpose language in order to gain certain specific evolutionary advantages. Language was the aural extension of tactile grooming. It conferred a more efficient way to build and sustain social bonds in an extended group of individuals who, together, were better at hunting and procreating. We still retain a great deal of that specific-purpose language today; modern human communication remains mostly about social gossip. We enjoy grooming each other with 'Likes' on Facebook and with mindless chitchat. Cognitive archaeology tells us that general-purpose language came rather late, and was perhaps what triggered the big bang of the modern mind. In other words, it was language that caused our brain to change from specific-purpose to general-purpose. Indeed, findings from modern neuroscience seem to provide evidence supporting this hypothesis. Our brain is a mixed bag. There are still many parts of our brain that are mysteriously focused and totally dedicated to something specific, say the recognition of faces. There are other parts that are more general-purpose; they connect 'dots' of information together in order to infer general things about the world. We are as yet at the very early stages of scientifically exploring the human brain and mind. But if these early discoveries offer any guidance, we should expect that the human brain is just another testament of how evolution works – not as the result of meticulous design but the haphazard addition or patching of new functionalities upon old ones.

General-purpose language is a similar hotchpotch product of circumstance. It is not concerned with precision but uses metaphor to convey information that is full of meaning and open to interpretation. Do aeroplanes actually 'fly', for instance; or do computers really 'think'? Is 'intelligence' the same thing as 'consciousness'? Is the brain a 'computer'? Unfortunately, we do not seem to care enough about answering these sorts of question properly nowadays. In our modern world of mass media and short attention spans, words are increasingly used as flashing slogans. You do not have to be a cunning marketing executive, or indeed a politician, to adopt a vocabulary that undermines the true meaning of words. Impressions have a higher monetisation value than reflections. But even if you are a well-meaning communicator pursuing the higher goal of true elucidation, language is firmly set against you. We discussed how the French post-structuralist philosopher Jean Baudrillard cast doubt upon whether the First Gulf War actually 'happened', by challenging the language and other communication devices used to communicate the war through mass media. Likewise, when we talk about 'Artificial Intelligence', 'consciousness' and the 'mind', we often lose ourselves in the translation of metaphors. And most narratives only add to the confusion by devising fictional characters that are machines with human characteristics, or souls. The conflicting literary narratives of love and fear condition the ways we discuss robots, androids and intelligent machines. But is there another, more precise and perhaps less poetic way to discuss Artificial Intelligence?

Trapped in metaphor

In *The Republic*, Plato proposes that poets should be banished from the *polis* and their works be destroyed.[3] He argues that poets create confusion by compressing too many meanings into words and expressions. As discussed earlier in this book, aeroplanes do not actually 'fly' but glide, powered by engines. Computers do not 'think', but process data by following logical algorithms. The words we use to describe what aeroplanes and computers do are metaphorical. For Plato, the use of metaphor – a creative tool when you are a poet (or a marketing executive, come to that) – is an abomination for the perfect *polis*. Language must be kept pure of metaphor in order to avoid confusion. Words must express one meaning only, so that the philosopher kings can always arrive at the indisputable truth. Conversations amongst citizens in a Platonic utopia must precisely reflect the definitions and concepts that exist in the world of ideals. Only when language is free from poetic ambivalence can people hope to be governed with justice and peace. For even the meaning of these two words – 'justice' and 'peace' – must be clearly defined for all to acknowledge. So is the solution to humanity's problems the banishment of literature, as Plato suggests?

Plato was not the only one frustrated with multiple meanings, wordplay and semantic confusion. We saw in Part II how Wittgenstein tried to 'purify' meaning in language through his seminal work, *Tractatus Logico-philosophicus*. In this book, the greatest philosopher of the twentieth century aimed to construct a logically perfect language. He assumed that meaning had to be objective, like things of the physical world, like the stuff scientists study in their labs. Similar to an atomic theorist, he also assumed that there had to exist 'logical atoms', snippets of perfect meaning that were pure and objective, and which one could use to construct sentences that would never be misinterpreted. After all, there could only be *one* truth. Wittgenstein felt so passionate about his conviction that he did his utmost to ensure that Russell and Whitehead succeeded in their epic attempt to discover the foundations of logic. He pored over their *Principia Mathematica*, and suggested corrections to help make the logic watertight. There could not be any contradictions, or paradoxes. Everything should be concluded from first principles by following logical rules. However, Gödel blew *Principia Mathematica* out of the water with his incompleteness theorem. After him, every notion that logic is the surest path to truth was shattered. This represented a turning point in the history of logic, science and philosophy that led Wittgenstein to reassess his original goal.

Wittgenstein's last book was published posthumously and was entitled *Philosophical Investigations.*[4] In this book he appears to contradict *Tractatus Logico-philosophicus.* He abandons the quest for 'logical atoms', the fundamental elements of meaning. Although he still advocates that most philosophical problems – for example, our ideas about the mind and whether it could be reproduced in a machine – are caused by conceptual confusions surrounding language, he now takes a completely different approach. He says that the meaning of any word does not come from a definition, but from its use via the word. Take, for example, the word 'game'. We all know what the word means, but it's very difficult to come up with a comprehensive definition of it. It is in fact impossible to devise rules that define the indisputable meaning of anything. There are 'games' for which winning or losing means very little to contestants, games of life and death – and so on. Each instance of meaning has its own rules. In other words, meaning cannot be captured in a 'formal logical system', such as the one aspired to in *Tractatus Logico-philosophicus.* Wittgenstein further acknowledges that we *do not need* to have an absolute definition of a word in order to use it. Meaning comes from the use of language, and it is therefore impossible to define meaning irrespective of words. Furthermore, Wittgenstein concludes, meaning is a social construct that happens between language users.[5]

This is an astounding insight that resonates perfectly with the findings of AI research. Although we can program a computer to process information and take autonomous actions, we cannot program one to understand meaning. This lack of understanding, according to Wittgenstein, cannot be remedied – ever. Computers will be always and forever unable to comprehend the meaning of the words we use (or which *they* use when communicating in natural language). And that is because computers are programmed using formal languages. These computer languages are different from the natural language we humans use to communicate. The programmer of a computer needs to define things *a priori*, in order for information processing to take place. We, as Wittgenstein showed, do not need definitions in order to use our language meaningfully.

Unfortunately, this huge difference between computers and humans has become a footnote in the contemporary debate about AI. Again, meaning and language confuse us. We are trapped in metaphor, because there is no other way to communicate amongst ourselves. Whatever Plato may have thought or desired, it is our poets and writers who define meaning through wordplay and the artistic use of narrative. We therefore think computers are intelligent not because they are, but because this is how natural language compels us to think.

As our society becomes ever more dependent on computers, our language will progressively assimilate their intrusion and will evolve new meanings around them. As computer technology progresses, and as Artificial Intelligence becomes embedded in information systems, computers will appear to be increasingly more 'intelligent'. We will increasingly have more personal and straightforward interactions with them, for example by talking to them in our own language. Intelligent computers – and intelligent 'things' in general – will become part of our everyday life and environment. From stand-alone objects, they will be transformed into social subjects, seamlessly connected to our human, social fabric. They will appear to understand us, and that would be good enough for us – for this will confirm how we have thought of artificial beings since prehistoric times. The Turing Test will be vindicated as the best method for testing Artificial Intelligence, because this is how we form our perceptions of and connections to other creatures, including other humans: on the basis of their behaviour. A computer that behaves intelligently *will* be considered intelligent, even if it is a philosophical zombie.

Moreover, because computers enhanced with Artificial Intelligence will have access to vast memory and processing resources, the computers of the future will be considered *more* intelligent than humans. But will we continue to trust them when they have outsmarted us? That will depend on what happens in the next few years. If computer systems become resistant to cyber attacks, and if nothing serious such as the Y2K or the 2010 flash crash happens, then our trust in intelligent computers will increase. Indeed, there may come a time when intelligent computers will be seen as the solution to all of humanity's problems, including how to better govern ourselves.

The end of liberty

Most of the ideas, or warnings, offered today about the future effects of Artificial Intelligence point to the labour market. Several economists, such as Tyler Cowen, have argued that AI will cost many white-collar jobs. However, their analysis assumes that all other things will remain more or less equal, for instance our political system of parliamentary representation, or our free economies of prices mostly regulated by markets. But this is not necessarily so. Indeed, history has already shown us that major technological changes are the causes of social and economic paradigm shifts. For instance, we refer to the invention of agriculture around 12,000 years ago as the 'agricultural revolution' because it completely changed how people lived and organised themselves. Nomads and hunter-gatherers who once roamed freely over lands belonging to no one became the subjects of kingdoms and empires with hereditary property laws. The Industrial Revolution that began in late eighteenth century created a new stratification in society, with the professional and entrepreneurial middle classes displacing the landed gentry and nobility. Land power was thus replaced by capital that could be moved freely, be invested anywhere, produce more and better products and make its owners a lot richer than ever dreamed of by the owners of land. As a reaction to capitalism, Marxism and socialism, as well as other totalitarian ideologies such as fascism and Nazism, proposed the equitable distribution of wealth. The tension between totalitarian utopias and free-market idealism resulted in the destruction of the ancient imperial order, caused the violent death of tens of millions during two world wars and accelerated the development of new technologies – of which computers were perhaps the most important one.

We now live at a time when the aftermath of the Cold War seems like a fading echo of the past. The apparent victory of liberal capitalism over communism, symbolised by the destruction of the Berlin Wall in November 1989, is nowadays doubted and challenged. The Great Recession that was set off in 2007 has demonstrated that unregulated financial markets create financial bubbles that can bring down the entire world economy. Millions of livelihoods have been destroyed in southern Europe, where double-digit unemployment has wiped out hope for the next two generations at least. Mistrust in the political establishment in Western liberal parliamentary democracies is at an all-time high. As the division of wealth between the rich and the rest widens, the public has started to feel that the game of capitalism is rigged strongly against them. Meanwhile, various forms of 'state capitalism' in China and other less free countries appears to be more successful in distributing wealth more equitably.

In this current global political milieu, the advent of intelligent computers with access to vast data will provide a new, revolutionary tool at the hands of governments. The ideological divide between free-market liberalism and government regulated socialism will begin to blur. The argument that free markets are better than central governments at managing the economy rests on the premise that governments cannot calculate effectively the myriad interconnecting parameters that contribute to an optimised economy. Therefore, when a government intervenes by regulating a market sector, it effectively chooses 'winners and losers' – inexorably becoming the cause of loss rather than benefit for the economy in question. There are numerous examples of government intervention having created more problems than solutions, the old Soviet economy being perhaps the most obvious one. But imagine a parallel universe in which the Soviets invented AI and developed a sophisticated data-based economy. Imagine Soviet intelligent supercomputers running simulations of every possible economic transaction in fractions of a second, and taking decisions about where production resources should be allocated in order to maximise the creation of economic value. Imagine those computers being able to make precise forecasts about the consequences of every possible course of economic action, and choosing the optimum one – not unlike what Deep Blue did when confronting Kasparov and beating him. In such a world, the outcome of the Cold War would have been completely different.

Of course, this parallel reality of victorious communism did not actually occur. But that does not mean we have seen the end of history. The future could belong to liberal democratic governments, which become the ultimate regulators of the economy by using intelligent supercomputers. There may indeed come a time when such governments will be indistinguishable from communist states. Centrally planned economies may return to the fore, given the game-changing combination of powerful AI technologies and big data. Such a scenario is not only possible but also probable, because the public expects, and demands, 'change'. And real change can only come from making economics an exact science by means of data and intelligent computers. We now have celebrated economists debating ad nauseam their opposing interpretations of economic events *after* these have happened.

In the future, we will have intelligent computers predicting what *will* happen *before* it occurs. 'AI economics' will rewrite every economics book. This major paradigm shift in economics will have profound and revolutionary repercussions in the role and power of governments. In other words, a future with Artificial Intelligence as the effective commander of national and international economies means the end of economic liberty and capitalism as we know it.

So much for society at large, but what about the individual? What will the effect of AI economics be on a personal level? One might say that our lives will become better, wealthier, healthier and more productive. AI systems will provide us with the best possible advice in whatever decision we need to make. They will run our homes, cars, bank accounts, investments and just about anything else that we currently spend so much time thinking about, often with very mixed results. Pure chance will be minimised, since AI systems will be able to explore almost every possible future scenario, and suggest the best course of action. AI will thus make us more successful – or at least deprive us of excuses for our failures. In other words, AI will prevent us from making mistakes. But is this *really* what we want?

Mistakes often result in very painful outcomes. For most of us, it would be great to avoid them thanks to the advice of our new mechanical friends. But mistakes also make us wiser. They turn us from children into adults. Their very possibility compels us to take responsibility for our actions. Mistakes also make us empathise with others who, human like we are, are also prone to erring. Until now, making mistakes and taking wrong decisions have been inescapable parts of life. The capacity for error characterises the human condition and informs our most cherished moral values, including charity and caring for others 'less fortunate' than ourselves. Yet, once we have the technology to help us make the best decisions most of the time, there will be no excuse for anyone failing. There will be no 'less fortunate' others, because chance will be virtually obliterated by an intelligent algorithm. The motivation to use such a technology will be enormous. If our lives are the sum of our decisions then, according to this line of thinking, living in a future with AI will be a great deal more rewarding than our existences today. Artificial Intelligence has the potential to make everyone reach perfection in their personal lives, by always choosing the right partner, profession, job – everything. Hard choices will become less hard. The unquantifiable that defines our moral lives will be quantified, because having the technological means to achieve maximum utility from our decisions will prove too great a temptation to ignore. Unchallenged by moral dilemmas, secure in the knowledge that we can do no wrong, we will be in danger of losing the most precious part of our humanity: our humility.

What a historical irony it would be if the intelligent machines that we created to be like us end up transforming us to become like them. Should this happen, humanity will find itself facing a daunting decision. Present-day post-humanism will have morphed into trans-humanism, the condition whereby humans adopt the elements, functions and characteristics of machines. Given the success of AI in improving everyone's life, as well as the proliferation of nanoscale computing devices embedded almost everywhere, there will doubtless be many who will aspire to become one with the machines, and integrate themselves with them. They might even be considered pioneers and the evangelists of progress. They will push society to create the next generation of truly conscious machines, a new cybernetic species that would be more than human.

But should society cross the line and bestow true life on such an artefact? The future of humanity will be defined by this dilemma.

Timeline:

A BRIEF HISTORY OF ARTIFICIAL INTELLIGENCE

65,000–40,000 years ago: Middle/Upper Palaeolithic transition. The big bang of the modern mind.

380 BC: Plato writes *The Republic*.

330 BC: Aristotle describes 'syllogisms', a way to mechanise thought.

150 BC: The Antikythera mechanism is constructed, a mechanical calculating machine, probably by students of astronomer Hipparchus of Rhodes.

AD 50: Hero of Alexandria designs first mechanical automata.

1275: Ramon Lull invents Ars Magna, a logical machine.

1637: Descartes declares *cogito ergo sum* ('I think therefore I am').

1642: Blaise Pascal invents the Pascaline, a mechanical calculator.

1726: Jonathan Swift publishes *Gulliver's Travels*, which includes the description of a machine that can write any book.

1801: Joseph Marie Jacquard invents a textiles loom that uses punched cards.

1811: Luddite movement in Great Britain against the automation of manual jobs.

1818: Mary Shelley publishes *Frankenstein*.

1835: Joseph Henry invents the electronic relay that allows electrical automation and switching.

1842: Charles Babbage lectures at the University of Turin, where he describes the Analytical Engine.

1843: Ada Lovelace writes the first computer program.

1847: George Boole invents symbolic and binary logic.

1876: Alexander Graham Bell invents the telephone.
1879: Thomas Edison invents the light bulb.
1879: Gottlob Frege invents predicate logic and calculus.
1910: Bertrand Russell and Alfred North Whitehead publish *Principia Mathematica*.
1917: Karel Capek coins the term 'robot' in his play *R.U.R.*
1921: Ludwig Wittgenstein publishes *Tractatus Logico-philosopicus.*
1931: Kurt Gödel publishes *The Incompleteness Theorem*.
1937: Alan Turing invents the 'Turing machine'.
1938: Claude Shannon demonstrates that symbolic logic can be implemented using electronic relays.
1941: Konrad Zuse constructs Z3, the first Turing-complete computer.
1942: Alan Turing and Claude Shannon work together at Bell Labs.
1943: Warren McCulloch and Walter Pitts demonstrate the equivalence between electronics and neurons.
1943: IBM funds the construction of Harvard Mark 1, the first program-controlled calculator.
1943: Charles Wynn-Williams and others create the computer Colossus at Bletchley Park.
1945: John von Neumann suggests a computer architecture whereby programs are stored in the memory.
1946: ENIAC, the first electronic general-purpose computer, is built.
1947: Invention of the transistor at Bell Labs.
1948: Norbert Wiener publishes *Cybernetics*.
1950: Alan Turing proposes the 'Turing Test'.
1950: Isaac Asimov publishes *I, Robot*.
1952: Alan Turing commits suicide with cyanide-laced apple.
1952: Herman Carr produces the first one-dimensional MRI image.
1953: Claude Shannon hires Marvin Minsky and John McCarthy at Bell Labs.
1953: Ludwig Wittgenstein's *Philosophical Investigations* published in German (two years after his death).

1956: The Dartmouth conference; the term 'Artificial Intelligence' is coined by John McCarthy.

1957: Allen Newell and Herbert Simon build the 'General Problem Solver'.

1958: John McCarthy creates LISP programming language.

1959: John McCarthy and Marvin Minsky establish AI lab at MIT.

1963: The US government awards $2.2 million to AI lab at MIT for machine-aided cognition.

1965: Hubert Dreyfus argues against the possibility of Artificial Intelligence.

1969: Stanley Kubrick introduces HAL in the film *2001: A Space Odyssey*.

1971: Leon Chua envisions the memristor.

1972: Alain Colmerauer develops Prolog programming language.

1973: The Lighthill report influences the British government to abandon research in AI.

1976: Hans Moravec builds the 'Stanford Cart', the first autonomous vehicle.

Early 1980s: The Internet is invented.

1982: The 5th Generation Computer Systems Project is launched by Japan.

1982: The film *Blade Runner* is released, directed by Ridley Scott, based on a short story by Philip K. Dick.

1989: Tim Berners-Lee invents the World Wide Web.

1990: Seiji Ogawa presents the first fMRI machine.

1993: Rodney Brooks and others start the MIT Cog Project, an attempt to build a humanoid robot child in five years.

1997: Deep Blue defeats Garry Kasparov at chess.

2000: Cynthia Breazeal at MIT describes Kismet, a robot with a face that simulates expressions.

2004: DARPA launches the Grand Challenge for autonomous vehicles.

2009: Google builds the self-driving car.

2011: IBM's Watson wins the TV game show *Jeopardy!*.

2014: Google buys UK company Deep Mind for $650 million.

2014: Eugene Goostman, a computer program that simulates a thirteen-year-old boy, passes the Turing Test.

2014: Estimated number of robots in the world reaches 8.6 million.[1]

2015: Estimated number of PCs in the world reaches two billion.[2]

NOTES

Introduction

1. PCs ('Personal computers') started becoming widely available in the early 1980s: IBM 5150 in 1981, Commodore PET in 1983. But at my university the first generation of PCs to be widely deployed for student use was IBM's legendary 'XT' that came out in 1988.
2. A 'neural correlate', as Crick called it.
3. According to German philosopher Karl Popper a theory is scientific only if it makes predictions that can be experimentally falsified. Testing a theory essentially means trying to find ways to disprove it.
4. Data according to the European Brain Council.
5. An annual lecture at the University of Cambridge named after Sir Robert Rede, Chief Justice of the Common Pleas in the sixteenth century.

PART I: DREAMING OF ELECTRIC SHEEP

1 The birth of the modern mind

1. See http://www.theartnewspaper.com/articles/Ice-Age-Lion-Man-is-worlds-earliest-figurative-sculpture/28595
2. Archaic *Homo sapiens* is dated to around 400,000 years ago (or earlier), although *Homo sapiens sapiens* appears in South Africa and the Near East around 100,000 years ago.
3. Evidence of artistic expression during the Middle Paleolithic (100,000–50,000 years ago) is scant and highly disputed.
4. Nevertheless, there have been cases with artistic chimpanzees. The most well-known case is Congo (1954–1964), a chimpanzee who learned how to draw and paint. Zoologist and surrealist painter Desmond Morris first observed his abilities when the chimp was offered a pencil and paper at two years of age. By the age of four, Congo had made 400 drawings and paintings. His style has been described as 'lyrical abstract impressionism'. This 'artistic' behaviour, evidently present only in very few individuals, and providing it is genuine, may show that the cognitive basis for

general intelligence was in place much earlier than currently assumed.

5 'Australopithecus' means 'southern ape'.

6 These tools are the products of what is called the 'Oldowan Industry', from the Olduvai Gorge in the Serengeti Plain, Tanzania. The most likely scenario is that *H. habilis* perfected these tools.

7 The *H. habilis* brain size was 500 to 800 cubic centimetres while the brain of the australopithecines *P. boisei* and *P. robustus* was between 400 and 500 cubic centimetres.

8 The 'Turkana boy'.

9 Fossils of archaic *Homo sapiens* have been found in Europe and have been dated 500,000 years ago. The species is called *H. heidelbergensis* and is supposed to have descended from *H. erectus*. The fossils have been found in Boxgrove, England, and Mauer, Germany.

10 Higham, T., et al. (2014), in: *Nature* (512), pp. 306–9.

11 Jones, J. S., and Rouhani, S. (1986), 'How small was the bottleneck?' in: *Nature* (319), pp. 449–50.

12 The Paleolithic is divided into three main periods, the Lower (*c.* 2,6Ma–300ka), the Middle (300–30ka) and the Upper (60–10ka). The big bang of the human mind occurred in the Upper Paleolithic.

13 Wynn, T. (1979), 'The intelligence of later Acheulian hominds', in: *Man* (14), pp. 371–91.

14 Theory of mind seems to be absent in many cases of autism, a very significant fact to which I shall return later in the book.

15 Humphrey, N. (1992), *A History of the Mind*. London: Chatto & Windus.

16 Dennett, D. (1991), *Consciousness Explained*. New York: Little, Brown

17 Lai, C. S. L., Fisher, S. E., Hurst, J. A., Vargha-Khadem F., and Monaco A. P. (2001). 'A forkhead-domain gene is mutated in a severe speech and language disorder', in: *Nature* 413, pp. 519–23.

18 Pinker, S. (1994), *The Language Instinct*. London: Allen Lane.

19 The biological determination of language is probably caused by many factors including many genes, and not just FOXP2.

20 Dunbar, R. I. M. (1993), 'Coevolution of neocortical size, group size and language in humans', in: *Behavioral and Brain Sciences* (16), pp. 681–735.

21 On 4 November 2013, I tried a simple experiment with Google Search. It returned 1.7 billion results for the keywords 'celebrity news' but a mere 49.7 million results for the keywords 'artificial intelligence'.

22 Talmy, L. (1988), 'Force dynamics in language and cognition', in: *Cognitive Science* (12), pp. 49–100.

23 Mithen S. (1996), *The Prehistory of Mind*. London: Thames & Hudson.

24 'In the beginning was the Word' (John 1:1). Perhaps one could liberally juxtapose the biblical term 'Word' (in Greek Λογος, which means reason, logic, but also speech) to the scientific term 'general-purpose language'.

2 Life in the Bush of Ghosts

1 Ingold, T. (1993), Tool-use, sociality and intelligence, in: *Tools, Language and Cognition in Human Evolution*, K. R. Gibson and T. Ingold (eds), pp. 429–45. Cambridge: Cambridge University Press.
2 Gellner, E. (1988), Plough, Sword and Book, in: *The Structure of Human History*. London: Collins Harvill.
3 After La Madeleine, a prehistoric site in south-western France.
4 Interestingly, modern neuroimaging research in altered states of mind shows that during ecstasy the limbic system of the brain takes over. It is as if our modern mind is disconnected and we re-experience the minds of our distant ape ancestors where the 'self' is dissolved and we feel 'part of the whole cosmos'.
5 Gazzaniga, M. S. (2011), *Who's in Charge? Free Will and the Science of the Brain*. Ecco: New York.

3 The mechanical Turk

1 The use and comprehension of metaphor is severely affected in people with autism, as several studies have shown.
2 Kuhn, T. (1979), Metaphor in Science, in: A. Ortony (ed.), *Metaphor and Thought*, pp. 409-19. Cambridge: Cambridge University Press.
3 Human, from the Latin *humus*, which means 'of the earth'.
4 Classical and Hellenistic Greece end at the Battle of Actium (31 BC) where Octavian beat the combined navies of Mark Antony and of the last Greek queen of Egypt, Cleopatra.
5 Galen's humourism theory of the physiology of the ciculatory system endured until 1628, when William Harvey published his treatise entitled *De motu cordis*, in which he established that blood circulates, with the heart acting as a pump.
6 With the exception of Homer who mentions the moving table in *The Iliad*.
7 Hephaestus was married to Aphrodite, the goddess of love.
8 The first part of ii, 15, of Constantinus Porphyrogenitus' (913–54) 'Book of Ceremonies' (*De Ceremoniis*) describes the reception in the Great Triclinium of the Magnaura Palace when a foreign legate visits Byzantium. The title describes it as taking place with the emperor seated ἐπί τοῦ Σολομωντείου θρόνου (on Solomon's throne). After

dealing with the preparation of the emperor and the court, it passes on to the ceremony, and here this sentence occurs: After the prime minister made his usual questions [towards the visiting dignitary] the lions and the birds started roaring, singing in harmony, and the animals on the throne started rising.

9 They are referred to as 'Roma-visaya'.

10 William of Malmesbury (*c.* 1125), in J. A. Giles (ed.), *William of Malmesbury's Chronicle of the kings of England; from the earliest period to the reign of King Stephen*. London: Henry Bohn, 1874 (p.174).

11 Dyson, G. (1999). *Darwin Among the Machines*. London: Penguin Books.

12 McCulloch, W. S. (1961), 'Where is Fancy Bred?', in: Henry W. Brosin (ed.), *Lectures in Experimental Psychiatry*. Pittsburgh: University of Pennsylvania Press.

13 IBM's Deep Blue beating Kasparov on 10 February 1996.

14 In fact he distinguished three substances, the third being God. Nevertheless, God also belongs to the non-physical, and that is why Descartes is considered the father of dualism.

15 Winter, A. (1998), *Mesmerised: Powers of Mind in Victorian Britain*. Chicago: University of Chicago Press.

16 Until a deeper level of reality is discovered; for example if string theory is shown to be experimentally correct.

17 Except in the United Kingdom and the British Empire where the Cooke and Wheatstone telegraph will persist.

18 Dyson, G. (1999), *Darwin Among the Machines*. London: Penguin Books.

19 McCorduck, P. (2004), *Machines Who Think: A Personal Inquiry into the History and Prospects of Artificial Intelligence*. New York: A. K. Peters.

20 'You shall not make for yourself an idol' under the Philonic division used by Hellenistic Jews, Greek Orthodox and Protestants except Lutherans.

21 Around the seventh century BC.

22 My discussion on the polar narratives about Artificial Intelligence focuses on Western civilisation; in Japan, for example, attitudes towards robots are not so polarised. Japanese society is more open to robots than Europeans and Americans.

4 Loving the alien

1 Turing, A. (1950), 'Computing Machinery and Intelligence', in: *Mind LIX* (236), pp. 433–466, ISSN 0026-4423.

2 'Gynaecoids' is a more appropriate term, from the Greek *gynē*: woman (while 'android' comes from the Greek *andro*: man). Sometimes the shortened term 'gynoid' is used, or 'fembot' (female robot).

3 Isaac Asimov introduced his 'three laws of robotics' in his 1942 short story 'Runaround'. The Three Laws are:
   1. A robot may not injure a human being or, through inaction, allow a human being to come to harm.
   2. A robot must obey the orders given to it by human beings, except where such orders would conflict with the First Law.
   3. A robot must protect its own existence as long as such protection does not conflict with the First or Second Laws.
4 Or the wall that separated the judge who fed the questions in Chinese and the people in the next room who processed an answer in Chinese by following a set of instructions.
5 *Blade Runner* is based on the 1968 Philip K. Dick novel *Do Androids Dream of Electric Sheep?*.
6 The original story of Pinocchio was conceived by Italian writer Carlo Collodi and published in 1883 as *The Adventures of Pinocchio*.
7 Freud used the word 'ambivalence' to note that love and hate could coexist for the same object.
8 Previously known as 'The Singularity Institute'.

5 Prometheus unbound

1 Title of the four-act lyrical drama by Percy Bysshe Shelley, published in 1820.
2 Tambora erupted in 1815 and was far more powerful than Krakatoa would be sixty-eight years later.
3 Grandfather of Charles Darwin.
4 Choreographed by Arthur Saint-Léon to the music of Leo Delibes, with libretto by Charles-Louis-Etienne Nuitter.
5 Frankenstein's monster refers to himself using these terms.
6 Shelley completed her writing in May 1817, and *Frankenstein; or, The Modern Prometheus* was first published on 1 January 1818 by the small London publishing house of Lackington, Hughes, Harding, Mavor, & Jones. The novel had been previously rejected by Percy Bysshe Shelley's publisher, Charles Ollier, and by Byron's publisher, John Murray.
7 The first AI system I ever developed, as an undergraduate student, was a medical expert system that diagnosed jaundice, an ailment of the liver. I had given that system the name 'Prometheus'.
8 Lucifer means the 'bearer of light'.
9 Based on the 1966 novel *Colossus* by Dennis Feltham Jones.
10 Iconoclasm in Greek means 'the breaking of icons'.
11 Saygin, A. P., Chaminade, T., Ishiguro, H., Driver, J., and Frith,

C., 'The thing that should not be: predictive coding and the uncanny valley in perceiving human and humanoid robot actions', in: *Social Cognitive and Affective Neuroscience* (2011); DOI: 10.1093/scan/nsr025.

12 Saygin's quote is taken from this article: http://www.utsandiego.com/news/2011/aug/04/ucsd-exploring-why-some-robots-creep-people-out/?sciquest

13 Blount, G., 'Dangerousness of patients with Capgras syndrome' (letter to editor), *Nebr Med J.* (7) 1: 207, 1986.

14 Ramachandran, V. S., and Blakeslee, S. (1999), *Phantoms in the Brain: Human Nature and the Architecture of the Mind*. London: Fourth Estate.

15 Using 'software' to describe the mind is a bad metaphor, as I will explain later. However, I am using this bad metaphor at this stage because it is culturally dominant and helps to clarify the point I am trying to make. In itself this is an example of how powerful metaphors (or scientific paradigms according to T. Kuhn) are, and how difficult it is not to use them.

6 The return of the gods

1 The Union of the Soviet Socialist Republics (USSR) ceased to exist on 26 December 1991.

2 Fukuyama, F. (1992), *The End of History and the Last Man*. London: Free Press.

3 Prominent structuralists include the French anthropologist Claude Lévi-Strauss, linguist Roman Jackobson and psychoanalyst Jacques Lacan. The father of structural linguistics, Ferdinand de Saussure, has exerted major influence on structuralists.

4 The list of post-structuralist philosophers includes philosopher and historian Michel Foucault, philosopher and social commentator Jacques Derrida, Marxist philosopher Louis Althusser, literary critic Roland Barthes, and philosopher Jean Baudrillard.

5 Latour B., and Woolgar S. (1979), *Laboratory Life: The Social Construction of Scientific Facts*. Los Angeles: Sage Publications.

6 Lyrics from the Beatles' song 'Strawberry Fields Forever'.

7 Baudrillard J. (1991), *La Guerre du Golfe n'a pas eu lieu*. Paris: Galilée.

8 I adopt the word 'metaphor' instead of Baudrillard's 'simulation' to avoid confusion with the same word, as I will use it to describe computer simulations of life and the brain in later chapters. I argue that my use of the word 'metaphor' in the context of discussing Baudrillard should be acceptable since Baudrillard's concept of 'simulation' correlates to a collective metaphor about reality

constructed by the human condition and human institutions (such as the media etc.).

9 Baudrillard, J. (1994), *Simulation and Simulacra*. Ann Arbor: University of Michigan Press.

10 Anything created using software (e.g. a digital image or a document) is a copy without an 'original'. Interestingly, fax machines may be considered society's initial – and last – resistance to creating digital simulacra. Although the document was created digitally using a word processor it was transformed into an analogue original that was then digitised anew, transmitted digitally, and reproduced analogously at the other end. The resistance failed because it attempted to retain an old mindset into the new technology of digital communications. Fax was replaced by the email. The advent of 3D printing renders physical objects copies without an original, just like email. Our world is increasingly populated by Baudrillardian simulacra. However, 'simulacra' are not only 'things' but ideas as well; for example, the idea of 'democracy'.

11 To further understand the difference let me introduce two philosophical terms. The first term is 'ontology' and it means knowledge about the essence, or nature of things. Ontology asks questions such as what are the fundamentals of being, or whether reality exists. If you hold an apple in your hand and ask yourself what is an apple, then you are asking an ontological question. The second term is 'epistemology' and is concerned with what knowledge is, and how we acquire it. Epistemology asks to what extent any given thing can be known. So with the apple in your hand you may now go ahead and ask how certain you can be that this is indeed an apple – you will be asking an epistemological question. Platonic ideas about reality are ontological questions – and these were the questions discussed in the film *The Matrix*. I will delve into these questions in the following chapters to explain them in depth, for they are fundamental in understanding Western philosophy and philosophy of mind. Nevertheless, Baudrillard's simulation argument is not ontological but epistemological. Arguably, structuralism and post-structuralism belong to epistemology.

12 The Wachowski brothers had given everyone on the set a copy of Baudrillard's book *Simulation and Simulacra* to read. In a scene in the first movie Neo is shown with the book, which he opens in order to hide his computer files.

13 Clynes, E. M., and Kline, N. S. (1960), 'Cyborgs and Space', in: *Astronautics*, 09/1960.

14 Based on the novel *The Cyborg* by Martin Caidin.
15 Terminator is often referred to as a cyborg. It is not. Terminator is an android.
16 Drugs may be considered cybernetic prostheses if their dose is constantly regulated by feedback signals from the patient's physiological measurements.
17 Hofstadter, D. (2007), *I Am a Strange Loop*. New York: Basic Books.

PART II: THE MIND PROBLEM

7 A blueprint for a universe

1 The original meaning of the Greek word 'tyrant' is a ruler who took rule by force and not by right of birth (the latter would be called a king, or *basileus* in Greek). It does not imply someone who is brutal or unjust. Plato was consultant to Dion of Syracuse and tutor to his nephew Dionysius II. Dion overthrew his brother and ruled Syracuse for a short time before being usurped by Calippus, a fellow disciple of Plato.
2 The Academy lasted for nearly a thousand years until it closed down by decree of Emperor Justinian in AD 529.
3 'Love with wisdom' in Greek is *philos* (friend) of *sophia* (wisdom), hence the word 'philosopher'.
4 Plato's *The Republic*, 473c-d.
5 Popper, K. R. (1971), *The Open Society and its Enemies*, Vol. 1, *The Spell of Plato*. Princeton: Princeton University Press.
6 Also known as 'Platonic solids': there are only 5 of these, with 4, 6, 8, 12 and 20 faces. They are the only regular, convex polyhedrons with regular polygons and the same number of faces meeting at each vertex.
7 And later the Cathars in the West.
8 Zarkadakis G. (2013), 'Ladder to heaven', in: *Aeon* magazine (26/03/2013).
9 The European Renaissance starts in the fifteenth and lasts till the seventeenth century. Scientific instrumentation has an older history that starts in the Middle Ages with the Alchemists.
10 The Eastern Greek Orthodox Church had separated from the Western Roman Catholic Church after the Great Schism in the eleventh century.
11 The big, yet-to-be-answered questions of science so far are, arguably, four: (a) why have we two descriptions of matter – gravity and quantum physics – and how can we unify them? (b) how did the

universe begin and why is it the way it is and not otherwise? (c) how did life begin? (d) how does the brain evoke consciousness?

12 There are four forces in nature according to modern physics; electromagnetism, the weak nuclear force, the strong nuclear force and gravity. So far the Standard Model unifies the first three forces. Gravity is the big unknown, hence M-theory or string theory.

13 In Timaeus.

14 Helmreich, S. (2000), *Silicon Second Nature*. Oakland: University of California Press.

15 A well-known example of a deformed protein is a prion. When it is folded in a certain way it is harmless. When it changes shape and deforms it becomes the cause of encephalopathy (mad cow disease). The word prion comes from the words protein and infectious.

16 Penrose, R. (1989), *The Emperor's New Mind*. Oxford: Oxford University Press.

17 Penrose, R. (1994), *Shadows of the Mind*. Oxford: Oxford University Press.

18 Spin networks are geometrical shapes in space–time that have the smallest possible dimensions of 10^{-33} cm and 10^{-43} seconds, the so-called 'Plank dimensions'.

19 Proven by the infamous Gödel theorem, to be fully discussed in Part III.

20 Quantum computers, however, would theoretically overcome this computational problem. As physicist David Deutsch has shown, quantum computers will demonstrate the existence of multiple universes. That is because quantum calculations will need far more states than there are particles in our universe, and will therefore have to take place 'outside' our universe.

8 Minds without bodies

1 Such shamanic rituals continue to today, for instance the Brazilian shamans who use a psychedelic brew called 'ayahuasca'.

2 Similar rituals were common throughout the Near East, for example the cults of Isis and Osiris in Egypt, the Persian mysteries, the worship of Adonis in Syria and the Cabeirian mysteries of Phrygia and Samothrace.

3 'I think therefore I am'.

4 Functional magnetic resonance imaging.

5 *La Monadologie*, 1714, is Gottfried Leibniz's most well-known works on metaphysics.

6 Eccles shared the Nobel Prize with Andrew Huxley and Alan Lloyd Hodgkin.
7 Eccles, J. C. (1994), *How the Self Controls Its Brain*. Berlin: Springer-Verlag.
8 Source: http://www.theguardian.com/science/2013/sep/21/stephen-hawking-brain-outside-body
9 Chalmers, D. (1996), *The Conscious Mind*. New York: Oxford University Press.
10 Nagel, T. (1974), 'What is it like to be a bat?', *The Philosophical Review*, Vol. 83, No. 4 (October), pp. 435–450.
11 Gould, S. J. (1997), 'Non-overlapping magisteria', *Natural History*, 106 (March), pp. 16–22.
12 The independence of the experimental apparatus from the observer ceased to be straightforward following the discovery of quantum physics. This is perhaps one of the reasons why quantum physics is often evoked in order to explain consciousness. There is a false logic at play stating that since quantum physics is a mystery and consciousness is a mystery, consciousness must have something to do with quantum physics. The relationship between the subjective and the objective in science is a lot more nuanced than that, as it will be examined later in the book.
13 Recent developments in epigenetics show that the genome, i.e. the way the base pairs are sequenced along the DNA molecule, is one of many factors that define hereditary characteristics. The environment seems to play a more important, and direct, role than previously thought.
14 Sets of base-pairs sequenced in a certain order that can pass hereditary characteristics (e.g. the production of a certain protein that builds an organ in an organism) are called 'genes'. The genome is the sequence of the genes in the DNA molecule. In the case of viruses the genome is coded in the RNA molecule (viruses do not have DNA).
15 The technique is used to explore the creation of new drugs and therapies, although it could also be used in bioterrorism.
16 According to the IMF the nominal GDP of our planet in 2012 was $71.7 trillion; agriculture was 5.9 per cent, industrial goods 30.5 per cent and services 63.6 per cent. I assume that most of the service sector is information processing in one way or another.
17 *Quis custodiet ipsos custodes*: the phrase is attributed to the Roman poet Juvenal from his *Satires* (Satire VI, lines 347–8), and is literally translated as 'Who will guard the guards themselves?' However, it is often linked to the worries of Socrates in Plato's *The Republic*

with regard to how the guardian class will be controlled. (Socrates suggests their moral training and proper cultivation of their souls.)

18 As documented in the Apostle's Creed recited by millions of Christians during Mass: 'I believe . . . in the resurrection of the body'.

19 Kurzweil, R. (2005), *The Singularity Is Near*. London: Viking.

20 Barrow, J. D., and Tippler, F. J. (1988), *The Anthropic Cosmological Principle*. Oxford: Oxford University Press.

21 Bostrom, N. (2003), 'Are you living in a computer simulation?', *Philosophical Quarterly*, Vol. 53, No. 211, pp. 243–55.

22 The idea of consciousness upload has been explored in several sci-fi novels and films, as for example in 'Kill Switch', an episode of the TV series *The X-Files* (written by cyberpunk pioneers William Gibson and Tom Maddox and aired in 1998). In the episode the main character uploads herself to the net.

23 Francis Crick has termed this hypothesis 'astonishing'; and uses it as the title of his seminal book on the neurobiological basis of consciousness. *The Astonishing Hypothesis* (1994). New York: Charles Scribner's Son.

9 La Résistance

1 See http://kepler.nasa.gov/

2 The Drake equation is: $N = R_* \cdot f_p \cdot n_e \cdot f_l \cdot f_i \cdot f_c \cdot L$ where:

N	=	the number of civilisations in our galaxy with which radio-communication might be possible (i.e. which are on our current past light cone);
R_*	=	the average rate of star formation in our galaxy;
f_p	=	the fraction of those stars that have planets;
n_e	=	the average number of planets that can potentially support life per star that has planets;
f_l	=	the fraction of planets that could support life that actually develop life at some point;
f_i	=	the fraction of planets with life that actually go on to develop intelligent life (civilisations);
f_c	=	the fraction of civilisations that develop a technology that releases detectable signs of their existence into space;
L	=	the length of time for which such civilisations release detectable signals into space.

3 Ward, P., and Brownlee, D. E. (2002), *Rare Earth: Why Complex Life is Uncommon in the Universe*. Berlin: Copernicus Books. Criticism of Drake's equation has come from many other scientists, too.

4 From the Italian physicist Enrico Fermi (1901–1954) who suggested it. Fermi was one of the fathers of nuclear energy and the atomic bomb.
5 Lindberg, D. C., Shank, M. H. (eds) (2013), *The Cambridge History of Science*, Vol. 2, *Medieval Science*. Cambridge: Cambridge University Press.
6 The set of natural numbers is made of all positive integers (1, 2, 3 ...). Since the nineteenth century zero (0) is also considered a natural number.
7 Aristotle's works are known by their Latin names. *De Anima* means 'On the Soul'; the original Greek title is *ΠερίΨυχής*.
8 As was the tradition at the Macedonian court King Philip placed his son Alexander together with two of his son's best friends, Cassander and Ptolemy, to study under Aristotle. Cassander and Ptolemy later became generals in Alexander's army and kings of Hellenistic kingdoms in Asia and Egypt respectively. Alexander, although initially in awe of Aristotle, became paranoid about him after he crushed the Persians, and suspected his former teacher of plotting to assassinate him.
9 Averroes is the Europeanised name of Ibn Rushd (1126–1198), an Andalusian polymath.
10 Avicenna is the Europeanised name of Ibn Sina (980–1037), the famous Persian philosopher.
11 To be fair, continental rationalists such as Descartes, Spinoza and Leibniz also advocated the empirical scientific method.
12 George Berkeley's ideas are usually called 'subjective idealism', a term suggesting that everything is in anyone's mind.
13 David Hume was the founder of scepticism.
14 From James Boswell's *The Life of Samuel Johnson* (1791).
15 Called the 'verification principle'.
16 Dennett, D. (1991), *Consciousness Explained*. New York: Little, Brown.
17 Chalmers, D. J. (2002), *Philosophy of Mind: Classical and Contemporary Readings*. Oxford University Press.
18 Indeed, Dennett's theory of consciousness is called 'Multiple Drafts model'.
19 Birds are different; they make a graphical map that relies on detection of angular movement by their retina.
20 Angular speed (or velocity) is how we measure the angular displacement of an object that moves around an axis. Instead of using as a unit 'centimetres' (which is a unit for something that moves in a straight line) per second we use 'degrees' per second. So 900^0 per second is really fast (just think that one full rotation is 360^0)!

21 Hayles, N. K. (1999), *How We Became Posthuman*. Chicago: University of Chicago Press.

22 Wiener, N. (1948), *Cybernetics*. Cambridge, Mass.: MIT Press.

23 Shannon was inspired by Marvin Minsky, the original inventor of the 'ultimate machine' and one of the fathers of Artificial Intelligence. Both Minsky and Shannon worked together at Bell Labs in 1952.

24 Perhaps the title of my book holds an 'encrypted' message. If you put the initials of the title together (IOOI), and transform them to binary numbers (1001), you get the decimal result 9 (whatever that means . . .)!

25 Shannon, C. E., and Weaver W. (1948), *The Mathematical Theory of Communication*. Champaign: University of Illinois Press. Shannon co-wrote the book with Warren Weaver, a pioneer in machine translation.

26 I am rephrasing here an example given by Katherine Hayles in her 1999 book, *How We Became Posthuman*.

27 The number of cells in our body is estimated between 5 billion and 200 billion. See: Bianconi, E., et al. (2013), 'An estimation of the number of cells in the human body', in: *Annals of Human Biology*, Nov–Dec 2013, Vol. 50, No. 6, pp. 463–71.

10 Peering into the mind

1 Miller, G. A. (1951), *Language and Communication*. New York: McGraw Hill.

2 Crick, F. (1995), *The Astonishing Hypothesis: The Scientific Search for the Soul*. New York: Scribner reprint edition.

3 Newberg, A. B., and Waldman, M. R. (2009). *How God Changes Your Brain: Breakthrough Findings from a Leading Neuroscientist*. New York: Ballantine Books.

4 Dehaene, S. (2014), *Consciousness and the Brain*. London: Viking.

5 I am making up this word 'noometer', from the Greek words *nous, νους* (mind) and *metron, μέτρον* (measure).

6 Ogawa S., Lee, T. M., Kay, A. R., and Tank, D. W (1990), 'Brain magnetic resonance imaging with contrast dependent on blood oxygenation', in: *Proc Natl Acad Sci*, Vol. 87, pp. 9868–72.

7 Baars, B. J. (1989), *A Cognitive Theory of Consciousness*. Cambridge: Cambridge University Press.

8 The Ig Nobel Prize is awarded each year for 'achievements that first make people laugh, and then make them think'.

9 Simmons, D., and Chabris, C. (1999), 'Gorillas in our midst:

sustained inattentional blindness for dynamic events', *Perception*, Vol. 28(9):1059–74.

10 You can do the experiment yourself by visiting www.theinvisible gorilla.com

11 Dehaene, S. (2014), *Consciousness and the Brain*. London: Viking.

12 Interesting experiments in sensory deprivation have taken place using isolation tanks (also called 'John C. Lilly tank', for their discoverer), lightless, soundproof tanks inside which subjects float in salt water at skin temperature. The idea of the tank was used in Ken Russell's 1908 film *Altered States*, and more recently in the sci-fi TV series *Fringe*.

13 Brain oscillations occur at different frequencies that include the 'alpha band' (8–13 hertz), the 'beta band' (13–30 hertz) and the 'gamma band' (30 hertz and higher). Francis Crick and his collaborator Christof Koch speculated that consciousness is reflected in gamma oscillations around 40 hertz (25 pulses per second) reflecting the circulation of information between the cortex and the thalamus. These finding have been revised. Dehaene's research has revealed that it is the late amplification of gamma band activity, rather than its mere presence, that constitutes the signature of conscious perception. Furthermore, it has been found that even an unconscious stimulus can induce high-frequency activity in the gamma band (40 hertz or higher).

14 Dehaene, S. (2014), *Consciousness and the Brain*. London: Viking.

15 CERN: The European Organisation for Nuclear Research, is a European research organisation based in Geneva whose purpose is to operate the world's largest particle physics laboratory.

16 ISS: the International Space Station.

17 ITER: the International Thermonuclear Experimental Reactor based in Cadarache, France.

18 BRAIN Initiative: Brain Research through Advancing Innovative Neurotechnologies.

19 HBP had thirty-two partners across thirteen countries on April 2014. Many more partners and countries are expected to join in due course.

20 Edelman, G. M., and Tononi, G. (2000), *A Universe of Consciousness: How Matter Becomes Imagination*. New York: Basic Books.

21 Tononi, G. (2012), *Phi: A Voyage from the Brain to the Soul*. New York: Pantheon Books.

22 Koch, C. (2012), 'A quest for consciousness', book review In: *Nature* (488), pp. 29–30.

23 Pert, C. et al. (1985), 'Neuropeptides and their receptors: a psychosomatic network', in: *Journal of Immunology*, Vol. 135(2), pp. 820–26.
24 Mørch, H., and Pedersen, B. K. (1995), 'Beta-endorphin and the immune system – possible role in autoimmune diseases', in: *Autoimmunity*, Vol. 21(3), pp. 161–71.

11 The cybernetic brain

1 From the word *κυβερνήτης* which means 'governor'. The word is used by Plato in *Alcibiades* to describe the act of governance of a people. In modern times it was first used by the French mathematician André-Marie Ampère who coined the word 'cybernetique' in his 1834 essay that described the science of civil government.
2 Stasis *(στάσις)* in Greek means to stay still but also means uprising, mutiny and upheaval.
3 In 1959 Frank Fremont-Smith, as head of the Macy Foundation, organised the first ever conference on LSD.
4 Wiener, N. (1948), *Cybernetics: Or Control and Communication in the Animal and the Machine*. Paris: Hermann & Cie, and Cambridge, Mass: MIT Press; 2nd revised edn 1961.
5 McCulloch, W. S., and Pitts, W. (1943), 'A logical calculus of the ideas immanent in nervous activity', in: *Bulletin of Mathematical Biophysics* (5), pp. 115–33.
6 Dehaene, N., and Naccache, L. (2001), 'Towards a cognitive neuroscience of consciousness: basic evidence of a workspace framework', in: *Cognition*, 79 (1–2), pp. 1-37.
7 Artificial Intelligence split from cybernetics in the summer of 1956 with its inaugural conference in Dartmouth, New Hampshire, one year before von Neumann's death.
8 Turing, A. M. (1936), 'On Computable Numbers, with an Application to the Entscheidungsproblem', *Proceedings of the London Mathematical Society*, 2 (1937), 42, pp. 230–65.
9 To be more accurate, Gödel encoded metamathematical statements within ordinary arithmetic.
10 The incomplete manuscript and notes based on a series of lectures given by von Neumann at the University of Illinois in 1949 was assembled and edited by Arthur Burks and published ten years after von Neumann's death.
11 New findings in epigenetics show that the mechanism of passing hereditary features to future generations is more complex than previously thought, and probably involves other systems in the cell beyond DNA replication.

12 The human mind may be beyond logical coding (as Gödel has indirectly showed) but it is not beyond computation. As we shall see, the deep connection between computation and life is the key to artificial life and artificial intelligence.
13 Hayles, N. K. (1999), *How We Became Posthuman*. Chicago: University of Chicago Press.
14 Rebeck, J. 'Synthetic Self-Replicating Molecules', *Scientific American*, 48–55 (July 1994).
15 Zarkadakis, G. (2001), 'Noetics: A proposal for a theoretical approach to consciousness', Proceedings of International Conference 'Toward a Science of Consciousness: Sweden 2001; Consciousness and its place in Nature', University of Skövde, Sweden, 7–11 August 2001.
16 Foerster, H. von (1981), *Observing Systems*. Seaside, Calif.: Intersystems Publications.
17 Hofstadter, D. (1979), *Gödel, Escher, Bach: An Eternal Golden Braid*. New York: Basic Books.
18 Hofstadter, D. (2007), *I Am a Strange Loop*. New York: Basic Books.
19 Corballis, M. C. (2014), *The Recursive Mind: The Origins of Human Language, Thought, and Civilisation*. Princeton: Princeton University Press.

PART III: ADA IN WONDERLAND

12 All Cretans are Liars

1 Aristotle's collective writings on logic were later grouped under a book entitled *Organon* (which means the 'instrument' in Greek).
2 More precisely, the title of George Boole's book was *An Investigation of the Laws of Thought on Which are Founded the Mathematical Theories of Logic and Probabilities*. It was the second of his two monographs on algebraic logic and was published in 1854.
3 In algebra we have six operations: addition, subtraction, multiplication, division, raising to an integer power and taking roots.
4 This is Euclid's theorem, which is a fundamental statement in number theory. It was proved by several mathematicians, including Leonhard Euler and Paul Erdös.
5 For example, Euclidian geometry accepts the improvable axiom that two parallel lines will never meet.
6 Formally this is regarded as the fourth property of a formal logical system, and is called 'soundness'.

7 In the post-modern literary context of the twenty-first century Swift's irony becomes a tenet: all texts can be regarded as self-produced, in that they have a transcendental-bibliographical animus which acts like a virus. Human minds are the hosts of this viral propagation, and mutation, of texts. The writer 'thinks' he is the creator but he is merely an empty vessel, a hapless idiot. Texts are virtually alive; they are 'memes'. As if history has come full circle, this notion is the same as that of the ancient Greeks and Romans.

8 Turing would publish his solution to the decision problem in 1936, reformulating Gödel's incompleteness theorem published in 1931.

9 The symbolic representation of the Barber's paradox can be programmed in a computer using Prolog, a language that was specifically designed for AI and uses Frege's predicate logic. In Prolog, Frege's notation would look like the following self-referencing clause:

```
Shaves (barber, X) : male(X), not shaves(X,X).
male (barber).
```

10 Whitehead and Russell were working on such a foundational level of mathematics and logic that it took them until page 86 of Volume II to prove that 1+1=2. They accompanied their proof with the dry comment: 'The above proposition is occasionally useful'.

11 Gödel, K. (1931), 'Über formal unentscheidbare Sätze der Principia Mathematica und verwandter Systeme, I', in: *Monatshefte für Mathematik und Physik.*, Vol. 38, pp. 173–98. (The title of the paper translates into English as: 'On Formally Undecided Propositions of Principia Mathematica and Related Systems'.)

12 Nagel, E., and Newman, J. R. (2001), *Gödel's Proof.* New York: New York University Press.

13 Epimenides' liar paradox can be restated as the following sentence: 'This statement is false'. That was the actual statement that Gödel 'formalised'.

14 To be more precise, Gödel proved that for any computable axiomatic system that is powerful enough to describe arithmetic of the natural numbers (e.g. the Peano axioms) the consistency of the axioms cannot be proven within the system.

15 Heisenberg's uncertainty principle essentially states that we can never know everything about a quantum phenomenon. Therefore, nature will remain forever at least partially unknown to us.

16 Turing, A. M. (1936), 'On Computable Numbers, with an Application to the Entscheidungsproblem', in: *Proceedings of the London Mathematical Society*, 1937, Vol. 2, No. 42, pp. 230–65.

17 The American mathematician Alonzo Church independently published his proof of the *Entscheidungsproblem* for the American Mathematical Society using a method called 'lamda calculus', and therefore the solution is known as the Turing-Church theorem. (Church, A. (1936), 'An unsolvable problem of elementary number theory', in: *American Journal of Mathematics*, 1936, Vol. 58, pp. 345–63.

18 Penrose, R. (1989), *The Emperor's New Mind: Concerning Computers, Minds and the Laws of Physics*. Oxford: Oxford University Press.

19 Or no output at all, if they never 'halt', as Turing showed.

20 Although Russell and Whitehead proved 1+1=2 is true in arithmetic. However, that proof does not preclude us humans from rethinking the problem from a completely different axiomatic set, and inventing a 'paradoxical arithmetic'. In effect, we can change the axioms at will – and that's a manifestation of intuitive, incomputable thinking.

21 Turing, A. M. (1939), 'Systems of Logic Based on Ordinals', in: *Proceedings of the London Mathematical Society*, pp. 161–228. The paper was based on Turing's 1938 PhD thesis of the same title (Princeton University).

22 Hofstadter, D. (1979), *Gödel, Escher, Bach: an Eternal Golden Braid*. New York: Basic Books.

13 The Program

1 Or 'memes', to use Richard Dawkins' paradigm, where ideas behave like genes and are governed by evolutionary forces.

2 Lord Stanhope (3rd Earl Stanhope and Viscount Mahon) was a scientist and statesman who worked on logic machines for several decades. His Demonstrator was a device capable of solving mechanically traditional syllogisms, numerical syllogisms and elementary probability problems.

3 Morris, I. (2010), *Why the West Rules–For Now: The Patterns of History, and What They Reveal About the Future*. New York: Farrar, Straus and Giroux.

4 These are the four measures that Ian Morris uses to construct his 'Human Social Development Index': energy capture, organisation, war-making capacity and information technology.

5 Babbage, C. (1835), *On the Economy of Machinery and Manufactures* (4th edition). London: Charles Knight.

6 What is also significant about recursive functions such as difference functions is that they describe complex, chaotic behaviours when the relation between the two variables computed is non-linear. When finite differences between the parameters of a difference

equation become infinitesimal, then the equations are called 'differential' and are fundamental to calculus.

7 Difference Engine No. 2 was reconstructed under Doron Swade, the then Curator of Computing at the London Science Museum. Reconstruction took place between 1989 and 1991, in order to celebrate the two-hundredth anniversary of Babbage's birth. In 2000, the printer, which Babbage originally designed for the difference engine, was also completed.

8 The Jacquard Loom was invented by French weaver and merchant Joseph Marie Charles, nicknamed Jacquard. It was based on an earlier invention by the master of automata, Jacques Vaucanson – of mechanical duck fame.

9 The next 'Turing-complete' machine after the Analytical Engine would be ENIAC, which was constructed in 1946.

10 Ada Lovelace called her notes simply 'Notes'.

11 The Bernoulli numbers are a sequence of rational numbers with many applications in engineering, and of great interest to mathematicians. For centuries mathematicians were fascinated by general ways to compute sums of integer powers (e.g. the sum of squares, or of cubes of the first *n* numbers). The quest for a general formula of computation persisted until the Swiss mathematician Jakob Bernoulli (1654–1705) 'discovered' a sequence of constants (the Bernoulli numbers) that provides a uniform formula for the sums of all powers.

12 Brynjolfsson E., McAfee A. (2014),*The Second Machine Engine*. New York: W.W. Norton & Co.

13 In Turing's description the tape with the symbols (the 'data') is separate from the table of instructions (the 'program'). In modern computers data and programs are stored in the same storage, a key insight that is part of the 'von Neumann architecture'.

14 According to historians Robert Friedel and Paul Israel at least twenty-two other inventors 'discovered' the incandescent lamp prior to Thomas Edison. However, it was Edison who developed the lamp into an effective source of electric lighting by selecting an effective incandescent material, achieving a higher vacuum and using a higher resistance filament.

15 Konrad Zuse invented the world's first programmable computer Z3, which became operational in May 1941.

14 From Bletchley Park to Google Campus

1. 'Global Information Report 2013', World Economic Forum (www.weforum.com).
2. This is a phrase from Greek philosopher Heraclitus (535–475 BC). The complete quote is: 'War is the father and king of all; some he has made gods, and some men; some slaves and some free. In Greek: Πόλεμος πάντων μὲν πατήρ ἐστι πάντων δὲ βασιλεύς, καὶ τοὺς μὲν θεοὺς ἔδειξε τοὺς δὲ ἀνθρώπους, τοὺς μὲν δούλους ἐποίησε τοὺς δὲ ἐλευθέρους.
3. In 1941 the *Daily Telegraph*, in secret collaboration with GC&CS, ran an innovative recruitment campaign in the form of a crossword competition. Winners were later approached for a position at Bletchley Park. Many years later Google would run a similar kind of recruitment campaign. Advertisements showed only the following: {first 10-digit prime found in consecutive digits of e}.com. Only those who understood what these numbers were, and found them, could then go to the 'secret' website, and apply for a position at Google.
4. With a significant contribution from Cambridge mathematician Gordon Welchman (1906–1985).
5. The most complex electronic device until then used only 150 valves.
6. ENIAC: Electronic Numerical Integrator And Computer.
7. von Neumann, J. (1945), 'First Draft of a Report on the EDVAC'.
8. Turing also produced a similar concept at around the same time, which he called 'Automatic Computing Engine' (ACE).
9. Officially called 'Strategic Defense Initiative' (SDI).
10. COBOL: Common Business-Oriented Language.
11. This is Jules Verne's 'lost novel', rediscovered by his great-grandson in 1989 and published in 1994.
12. Another interesting piece of fiction is the short story 'The Machine Stops' written by E. M. Forster in 1909, in which communication is through an instant messaging/video conference machine.
13. Wright, A. (2014), *Cataloging the world: Paul Otlet and the Birth of the Information Age*. Oxford: Oxford University.
14. A conference room at Google's Brussels office has been given the name 'Mundaneum' in honor of Paul Otlet.
15. ARPANET: Advanced Research Projects Agency Network.
16. According to its inventor, the Web 'runs on the Internet like a fridge uses the power grid'. Quote taken from an interview with Sir Berners-Lee by Andrew Edgecliffe-Johnson, published on 7 September 2012 in the *Financial Times*.

17 HTML: Hypertext Markup Language.

18 Borges, J. L. (1941), *El jardin de senderos que se bifurcan*. Buenos Aires: Editorial Sur.

19 Negroponte, N. (1995), *Being Digital*. New York: Alfred Knopf.

20 Moore, G. E. (1965), 'Cramming more components on to integrated circuits', in: *Electronics Magazine* (4).

21 Called 'Digital Alpha 21164' microprocessor, developed and fabricated by DEC (Digital Equipment Corporation).

22 Electrons, being elementary particles, can penetrate through physical barriers thanks to 'quantum tunnelling'. This phenomenon is significant in very small scales, when electrons begin to act as waves. As we miniaturise electronics, this phenomenon becomes significant. At small scales we cannot assume that electrons will behave 'normally' as particles – and therefore we cannot be certain of their behaviour: this means that our designs fail.

23 Krauss, L. M., and Starkman, G. D., 'Universal Limits of Computation', 10 May 2004. http://arXiv:astro-ph/0404510

24 One zettabyte equals 1,000 exabytes. One exabyte amounts to 36,000 years of HD-TV video, or the equivalent of streaming the entire Netfliz catalogue 3,177 times.

25 VNI Forecast highlights, Cisco, cisco.com/web/solutions/sp/vni/vni_forecast_highlights/index.html

26 Infographic, 'The Dawn of the Zettabyte Era', Cisco Blogs, blogs.cisco,com/news/the-dawn-of-the-zettabyte-era-inforgaphic/

27 Y2K: Overhyped and oversold? BBC News Report, 6 January 2000.

28 Lewis, M. (2014), *Flash Boys: Cracking the Money Code*. London: Allen Lane.

29 Duhigg, C. (16 February 2012), 'How companies learn your secrets', in: *The New York Times Magazine*.

30 Miller, M. (2014) 'Cisco Predicts "Internet of Everything" Explosion', PC magazine on-line, accessed 16 July 2014.

15 Machines that think

1 Brynjolfsson, E., and McAfee, A. (2014), *The Second Machine Age: Work, Progress, and Prosperity in a Time of Brilliant Technologies*. New York: W. W. Norton & Co.

2 In FY 2013 figures, Google ranks second amongst its peers in terms of market capitalisation (Apple is first), but tenth in terms of revenues (where Samsung is first) – data retrieved from Wikipedia on July 2014.

3 John McCarthy would later set up the first AI lab at MIT with Minsky.

4. Rochester designed the IBM 701, the first general purpose, mass-produced computer.
5. The epic film *2001: A Space Odyssey* was produced and directed by Stanley Kubrick, who co-wrote the screenplay with Arthur C. Clarke.
6. For instance, the 'Logic Theorist' built in 1955 and 1956 by Allen Newell, Herbert Simon and John Shaw, a system that solved logical problems.
7. An attempt to go beyond the 'yes'/'no' bifurcation in AI was made by using 'fuzzy logic'. In fuzzy logic something can be of a certain kind, or belong to a certain category, only partially. Nevertheless, the problem of general reasoning has not being solved by fuzzy logic either.
8. The early years of AI, between 1956 and 1974, are also known as the 'golden period' of AI.
9. The Japanese Ministry of International Trade and Industry earmarked $850 million to the 5th Generation Computer Project. It was abandoned by the end of the 1980s.
10. The name MYCIN was derived from -mycin, the suffix accompanying many antibiotics.
11. Zarkadakis G., Carson E. R., Cramp D. G., and Finkelstein L. (1989), 'ANABEL: Intelligent Blood-Gas Analysis in the Intensive Care Unit', in: *International Journal of Clinical Monitoring and Computing*, Vol. 6, pp. 167–71.
12. Deep Blue beat Kasparov in the second six-game match, 2–1 with 3 draws.
13. In 1997, Deep Blue ranked as the 259th most powerful supercomputer in the world.
14. Silver, N. (2012), *The Signal and the Noise*, London: Penguin.
15. Frey, C. B., and Osborne, M. A. (2013), 'The Future of Employment: How susceptible are jobs to computerisation?', in: *Machines and Employment Workshop*, Oxford: Oxford University Engineering Science Department and Oxford Martin Programme.
16. Cowen, T. (2013), *Average is Over: Powering American beyond the Age of the Great Stagnation*. New York: Dutton.
17. Brynjolfsson, E., and McAfee, A. (2014), *The Second Machine Age: Work, Progress, and Prosperity in a time of Brilliant Technologies*, New York: W.W. Norton & Co, p. 132.
18. Piketty, T. (2014), *Capital in the 21st Century*. Cambridge MA: Harvard University Press.
19. Krugman, P. (2013), 'Sympathy for the Luddites', in: *New York Times*, 13 June 2013.

20 Bostrom, N. (2014), *Superintelligence: Paths, Dangers, Strategies*. Oxford: Oxford University Press.
21 Tegmark, M. (2014), 'Humanity in Jeopardy', in: *Huffington Post*, 13 January 2014.
22 Tegmark, N., Hawking, S., Russell, S., and Wilczek, F. (2014), 'Transcendence looks at the implications of artificial intelligence – but are we taking AI seriously enough?', in: *Independent*, 1 May 2014.
23 Vinge, V. (1993), 'The coming technological singularity: how to survive in the post-human era', presented at: the VISION-21 Symposium sponsored by NASA Lewis Research Center and the Ohio Aerospace Institute, 30–31 March, 1993.
24 Vinge, V. (1993), 'The coming technological singularity: how to survive in the post-human era', presented at: the VISION-21 Symposium sponsored by NASA Lewis Research Center and the Ohio Aerospace Institute, 30–31 March, 1993.
25 Kurzweil, R. (1999), *The Age of Spiritual Machines*, New York: Viking Penguin.
26 Dowe, D. L., and Herandez-Oralli, J. (2011), 'IQ tests are not for machines, yet', in: *Intelligence*, March–April 2012, Vol. 40, No. 2, pp. 77–81.
27 Although Earth, as a cybernetic system, acts in an 'intelligent' way through constant adaptation and self-regulation, a concept explored by James Lovelock in his Gaia hypothesis.
28 Vinge, V. (1993), 'The coming technological singularity: how to survive in the post-human era', presented at: the VISION-21 Symposium sponsored by NASA Lewis Research Center and the Ohio Aerospace Institute, 30–31 March, 1993.
29 At the time this book was written in 2014 the fastest computer in the world was the Chinese Tianhe-2, located at Sun Yatsen University, Guangzhou, China. It could perform 33.86 petaflops, or 33.86 quadrillion floating point operations per second.
30 Moravec, H. (1988), *Mind Children*. Cambridge MA: Harvard University Press, p. 15.
31 Pinker, S. (1994), *The Language Instinct*. London: Harper Perennial.
32 Brooks, R. (2002), *Flesh and Machines*. New York: Pantheon Books.
33 Fitzgerald, F. S. (1945), *The Crack-up*. New York: New Directions.
34 TED talk by Ruth Chang, filmed May 2014, 'How to make hard choices', searchable on www.ted.com
35 Dreyfus, H. (1971), *What Computers Can't Do*. New York: MIT Press.

16 Darwin at the edge of chaos

1 Also called 'immunoglobulins'.

2 Chua, L. O. (1971), 'Memristor—The Missing Circuit Element', *IEEE Transactions on Circuit Theory*, Vol. CT-18, No. 5, pp. 507–19.

3 Kim, K. H. *et al*, (2012), 'A functional hybrid memristor crossbar-array/CMOS system for data storage and neuromorphic applications', in: *Nano Letters*, Vol. 12, No. 1, pp. 389–95.

4 Complementary metal-oxide-semiconductor (CMOS) is a technology used to manufacture microelectronics.

5 Pickett, M. D., Medeiros-Ribeiro, G., and Williams, R. S. (2013), 'A scalable neuristor built with Mott memristors', in: *Nature Materials*, Vol. 12, pp. 114–17.

6 The number 1,024 is very special in computing. It is 2^{10}, and is represented in binary as the simple round number 10000000000. It is often used in place of 1,000 as the quotient of a byte, kilobyte, megabyte, etc.

7 Rubenstein, M., Cornejo, A., and Nagpal, R., (2014), 'Programmable self-assembly in a thousand-robot swarm', in: *Science*, Vol. 345, pp. 795–99.

8 Brooks, R. (1999), *Cambrian Intelligence: the early history of the new AI*. New York: MIT Press.

9 Drexler, K. E. (1986), *Engines of creation: The coming era of nanotechnology*. New York: Doubleday.

10 The use of watches and clocks as examples of self-reproducing automata was apparently quite common in the eighteen and early nineteenth centuries. When Descartes went to work as tutor to young Queen Christina of Sweden, it is said that his royal pupil asked him to explain the workings of the human body. Descartes answered that the body could be regarded as a machine; whereby the Queen pointed at a clock on the wall, and ordered him to 'see to it that it produces offspring'.

11 The article was published on 13 June 1863, with Butler signing himself as 'Cellarius'.

12 Jones, R., Haufe, P., Sells, E., Iravani, P., Olliver, V., Palmer, C., and Bowyer, A. (2011). 'Reprap–the replicating rapid prototyper', in: *Robotica*, Vol. 29, No. 1, pp. 177–91.

13 Dyson, F. J. (1970), 'The Twenty-first Century', Vanuxem Lecture delivered at Princeton University, 26 February 1970.

14 Mora, T., and Bialek, W. (2011), 'Are biological systems poised at criticality?', in: *Journal of Statistical Physics*, Vol. 144, pp. 268–302.

15 Lettvin, J. Y., Maturana, H. R., McCulloch, W. S., and Pitts, W. H.

(1959), 'What the frog's eye tells the frog's brain', in: *Proceedings of the Institute of Radio Engineers*, Vol. 47, pp. 1940–51.

16 Autopoiesis comes from the Greek, meaning 'self-creation'.

17 Maturana, H., and Varela, F. (1988), *The Tree of Knowledge*. London: Shambhala.

18 Swenson, R. (1992), 'Galileo, Babel and Autopoiesis (It's turtles all the way down)', in: *International Journal of General Systems*, Vol. 21, No. 2, pp. 267–69.

19 Kauffman, S. (1995). *At Home in the Universe: The Search for the Laws of Self-Organization and Complexity*. Oxford: Oxford University Press.

20 Rebeck, J. (1994), 'Synthetic Self-Replicating Molecules', in: *Scientific American*, July 1994, pp. 48–55.

21 Wolfram had conjectured that Rule 110 was Turing complete. Proof that Rule 110 is Turing complete was published by Matthew Cook. The proof is found in: Wolfram, S. (2002), *A New Kind of Science*, Wolfram Media.

22 Or a transition to another state of dynamic equilibrium. In this case we have a 'bifurcation' point beyond which the system transits unpredictably to another attractor. An interesting speculation about such a mechanism applied in evolution was made by the late evolutionary biologist Stephen Jay Gould (1941–2002). His 'punctuated equilibria hypothesis' is a description of abrupt phase transitions on the scale of an ecosystem.

23 There exist 88 possible unique elementary cellular automata, but only Rule 110 has been proven to be Turing complete so far. It is possible that other unique cellular automata are also Turing complete, but more complex.

24 Baron-Cohen, S. (1997), *Mindblindness: an essay on autism and theory of mind*, New York: MIT Press.

25 Willingham, E. (2012), 'Is Autism an "Epidemic" or are we just noticing more people who have it?', in: *Discover* on line blog, 11 July 2012.

26 Hayles, N. K. (1999), *How We Became Posthuman*. Chicago: University of Chicago Press, p. 148.

27 Baron-Cohen, S. (1997), *Mindblindness: an essay on autism and theory of mind*, New York: MIT Press.

28 In April 1932, at the Cavendish Laboratory in Cambridge, UK, John Cockcroft and Ernest Walton split the atom for the first time.

Epilogue: The future of humanity

1 Findings of cave paintings in Indonesia challenge the Eurocentric view of the first art. See: Aubert, M., et al., Pleistocene cave art from Sulawesi, Indonesia (2014), in: *Nature* (514), pp. 223–7. There may indeed be other places around the world where art was created at the same time, or even earlier, as our species left Africa and colonised the world.

2 Damasio, A. (1999), *The Feeling of What Happens: Body and Emotion in the Making of Consciousness*. New York: Harcourt.

3 He took exception to Homer only.

4 Wittgenstein, L. (1999), *Philosophical Investigations*. Upper Saddle River: Prentice Hall.

5 Wittgenstein, with his two seemingly contradictory works, straddles the divide between Platonism and Aristotelianism. In his early work he aligns with Plato and tries to expel poets from the perfect *polis*: he believes that there is a world of ideals in which logical simples exist, and which, once discovered, the construction of a perfect language – one that cannot be misinterpreted by anyone – is possible. This perfect, Platonic language is the object of the *Tractatus Logico-philosophicus*. Nevertheless, with his last book Wittgenstein takes an empirical position: meaning is embedded in the use of language, it does not – and cannot – exist outside a social context. Meaning happens only when we communicate with another person who understands our language.

Timeline: A brief history of Artificial Intelligence

1 According to: http://spectrum.ieee.org/automaton/robotics/industrial-robots/041410-world-robot-population

2 According to research by Forrester Group.

ACKNOWLEDGEMENTS

My journey in Artificial Intelligence and the scientific study of consciousness began almost three decades ago. Throughout this time I was fortunate to encounter many brilliant minds that influenced my thinking and opened new vistas of enquiry. It would be impossible to list all of them, but I would like to acknowledge in particular the contribution of my research supervisor, Ewart Carson, at City University, London and Janos Sztipanovits at Vanderbilt University, Nashville, where I validated my expert system and learned to love America; as well as acknowledge those whom I met at the Consciousness Conferences in Tucson, including Stuart Hameroff, David Chalmers, Daniel Dennett, Steven Pinker, John Searle and Christof Koch.

In my years in Athens I was invited to participate in a very special circle of discussions about the mind at Athens University led by neuropsychologist Andrew Papanicolaou (currently based at the University of Tennessee). The circle included psychiatrists, clinical psychologists, philosophers of science and neuroscientists. It felt as if I were back in classical times – with the added advantage of hard scientific facts; a very enjoyable and productive experience indeed, for which I remain most grateful.

Special thanks are due to the British Council for trusting me as their collaborator in Famelab International, the international competition for young science communicators. They gave me the opportunity to work with young, enquiring, talented scientists and engineers from all around the world, who are passionate about sharing science with everyone. We are all on this planet together, and these young people from every continent, religion and culture are the living testament of this ideal. Their enthusiasm for, and dedication to, solving the big and small problems of science, and engaging with society at large, gives me hope for the future of humanity. Interacting with these beautiful young minds has helped me become a better science communicator.

I would also like to thank Stephen Webster at the Science Communication Group of Imperial College, London, for giving me the opportunity of a one-year research fellowship that included access to the University's library. Much of the research for this book was made possible because of this.

This book would have been simply impossible without my editor, Sue Lascelles. She helped me to clarify my thoughts, focus on the important things, and to think of my book as being more like a dialogue than a monologue. Writing a book often feels like climbing up a steep mountain whose top is clouded over, into unchartered and invisible territory. I was very lucky to have had Sue as my guide.

Finally, I would like to thank my wife, Victoria, and my son, Konstantinos-Nikephoros, for shining the light of love into my life, and for making me care enough to want to write.

INDEX